纺织服装高等教育"十二五"部委级规划教材
纺织工程专业双语教材

Woven Structure and Design
织物结构与设计

聂建斌　卢士艳　主编
陈晓钢　审

东华大学出版社

内容简介

本书是为高校纺织专业双语教学需要而编写的,其对应的中文教材为《织物结构与设计》。

本书加入了大量的新知识、新技术,也可作为纺织工程技术人员开发新产品的参考。

本书在书页的边部加入了部分专业词汇的中文注释,以帮助读者提高阅读效率,并正确把握专业词汇含义。

图书在版编目(CIP)数据

织物结构与设计/聂建斌,卢士艳主编. —上海:东华大学出版社,2014.6

双语教材

ISBN 978-7-5669-0521-5

Ⅰ.①织… Ⅱ.①聂…②卢… Ⅲ.①织物结构—双语教学—高等学校—教材②织物—设计—双语教学—高等学校—教材 Ⅳ.①TS105.1

中国版本图书馆 CIP 数据核字(2014)第 102784 号

责任编辑:库东方
封面设计:魏依东

织物结构与设计

聂建斌　卢士艳　主编　　陈晓钢　审
东华大学出版社出版
上海市延安西路1882号
邮政编码:200051　　电话:(021)62193056
新华书店上海发行所发行　　昆山市亭林印刷有限责任公司印刷
开本:787mm×1092mm　1/16　印张:16　字数:462千字
2014年6月第1版　2014年6月第1次印刷
ISBN 978-7-5669-0521-5/TS · 489
定价:39.00元

Perface
前 言

Document 4(2001) from China Ministry of Education requires that dual-language teaching should be practiced in some programs in all universities. This would obviously help Chinese university students to improve their English ability, and this policy has been considered as an effective measure taken towards the production of high quality international professionals. As to textile industry in our country, it is one of the most benefited industries from the open-door polices and is now entering global market in an impressive manner. Naturally, more professionals are expected urgently in the textile industry who are able to use English fluently to deal with technological and trade issues. This textbook is written for dual-language teaching for university students, and as a general reading reference.

This textbook includes all contents in the Chinese textbook *Fabric Structures and Design*, and has added some new information such as multi-layer fabrics. This textbook received satisfactory remarks from some specialists and students during the trial period.

The authors worked at their best to ensure the correctness in English grammar and language custom. Dr. Xiaogang Chen, senior lecture and Ph. D student supervisor of UMIST, UK was invited to be the auditor of this textbook and he has made corrections and examined the contents of the textbook.

The authors wish to thank the leadership of Zhongyuan Technology Institute, especially professor Delin Ling, the former president of the Institute, for their strong support and encouragement.

Comments and recommendations from readers are welcome.

国家教育部教高[2001]4号文件要求各大学在每个学科都有一些课程采用双语教学,这对提高我国大学生的英语使用能力很有帮助,对培养国际型高素质人才是一个非常好的举措,尤其是我们纺织行业,正在大步走向世界,从技术角度、贸易的角度都需要大量的、能熟练掌握英语的专门型人才,本书正是为高校纺织专业学生编写的双语教学教材,也可作为专业英语教材与读物。

本书包含《织物结构与设计》的所有内容,并增加了一些新的知识,如多层织物等,在试用中得到了专家和学生们的好评。

本书在语言上力求准确,符合英语习惯。英国UMIST大学纺织系陈晓钢博士,对本书进行了严格的修改和审查。

在编写本书过程中,得到了中原工学院领导、特别是老院长凌德麟教授的大力支持,在此表示诚挚的谢意。

热忱欢迎读者对本书提出宝贵意见和建议。

Introduction of the Author
作者简介

Mr. Jianbin Nie is currently a professor in School of Textiles, Zhongyuan University of Technology. He is also a National Excellent Teacher. He graduated from Northwest Institute of Textile Science and Technology(Xi'an) in 1982. He was a fabric designer in Beijing Wool Textile Research Institute from 1982 to 1993, and a senior fabric designer in Henan Textile Research Institute. He became an associate professor in Zhongyuan Institute of Technology in 1996.

Mr. Nie is a qualified fabric designer, having 30 years experience in this field. He has designed thousands of fabrics varieties. He participated a "7th five-year plan" national program, on "Development of Woolen Fabrics using Synthetic Fibers". He was one of the main investigators for the project "Development of Woolen Fabrics from Wool and Flax Blend" sponsored by Beijing Science Council, and awarded the second prize for scientific progress. He was an advisor for dozens of companies.

Mr. Nie is good at English due to his opportunities of English training and abroad experience. He had an English training at Beijing Second Foreign Language Institute from 1983 to 1984. He studied at UMIST U.K., from 1999 to 2000, and he worked as an expert at Bahir Dar University, Ethiopia, from 2001 to 2003.

WOVEN STUCTURE AND DESIGN, suitable for dual-language teaching, is a fruit of author's design skills, English ability and teaching experiences. It is convinced that this book will give a useful contribution to the dual-language teaching program.

聂建斌,中原工学院纺织学院教授,全国优秀教师,1982年毕业于西北纺织工学院,1982—1993年在北京毛纺织研究所产品研究室工作,1993—1996年在河南省纺织研究院工作,1996年至今在中院工学院纺织学院做教学工作。

聂建斌先生擅长于纺织品设计与开发工作,在长达30年的设计生涯中,他设计开发了千余种纺织新产品,并参加了国家"七五"攻关科研项目——"纯化纤毛呢产品的开发",主持了北京市科委的课题"毛麻产品的开发",并获科技进步二等奖。他还曾兼任二十余家工厂的技术顾问。

聂建斌先生具有丰富的学习和使用英语的经验,除在大学的学习外,1983—1984年在北京第二外国语学院进修一年,1999—2000年在英国UMIST大学纺织系留学一年,2001—2003的在埃塞俄比亚Bahir Dar大学纺织系任教(英语讲授纺织品设计等课程)。

《机织物组织与设计》是一本适合于双语教学的教材,集作者扎实的专业知识、较深的英语功底和丰富的教学经验于一体,相信它在中国的双语教学改革中,一定能发挥较大的作用。

Introduction to the Reader
审稿者简介

Dr. Xiaogang Chen is currently a professor in School of Material, University of Manchester, U.K. He received his BSc and MSc degrees in Textile Engineering from Northwest Institute of Textile Science and Technology (Xi'an) in 1982 and 1985 respectively, and obtained his PhD degree in Textile Engineering from University of Leeds, U.K. in 1991. After 3 years post-doctoral research in Heriot-Watt University, he joined UMIST as a lecture in 1994 and became a senior lecturer in 2002. UMIST forms part of the new University of Manchester from October 2004.

Dr. Chen's research interests include protective textiles, 3D technical textiles, modeling of textile assemblies and CAD/CAM for textiles. Over the years, he has been the principal investigator for a number of research projects supervised numerous PhD, MPhil, and MSc students.

His E-mail address is xiaogang. chen @umist. ac. uk.

陈晓钢博士,英国曼彻斯特大学材料学院教授,博士生导师,1982年和1985年在西北纺织工学院取得学士和硕士学位,1991年在英国里兹大学取得博士学位,在赫瑞—瓦特大学做了三年博士后研究,1994年开始任教于 UMIST 大学纺织系,2002年升级为高级讲师。UMIST 于 2004 年 10 月成为新曼彻斯特大学。

陈博士的研究领域包括防护纺织品、三维技术纺织品、纺织材料几何和力学建模,以及纺织 CAD/CAM。多年来,陈博士在多个领域指导了一大批硕士和博士研究生。

CONTENTS/目录

UNIT Ⅰ General Knowledge and Simple Construction
基本知识和简单组织 ··· 1

Chapter One General Knowledge on Woven Fabrics / 有关机织物的基本知识 ············ 2
1.1　Cloth formation on loom / 机织物的形成 ································ 2
1.2　Fabric analysis / 样品分析 ··· 4
1.3　Fabric representation / 织物的表示方法 ·································· 8

Chapter Two Elementary Weaves (Fundamental Weaves) / 三原组织 ············ 21
2.1　General characteristics / 基本特征 ····································· 21
2.2　Plain weave / 平纹组织 ·· 21
2.3　Twill weaves / 斜纹组织 ··· 23
2.4　Sateen / Satin weaves / 缎纹组织 ······································ 26

Chapter Three Derivatives of Elementary Weaves / 变化组织 ················· 29
3.1　Plain weave derivatives / 平纹变化组织 ································· 29
3.2　Twill weave derivatives / 斜纹变化组织 ································· 33
3.3　Derivatives from satin / sateen weaves / 缎纹变化组织 ···················· 51

Chapter Four Combined Weaves / 联合组织 ··································· 56
4.1　Stripe and check weaves / 条格组织 ····································· 56
4.2　Crepe weaves / 绉组织 ··· 61
4.3　Mock leno weaves / 假纱罗组织 ··· 68
4.4　Huckaback weaves / 浮松组织 ··· 71
4.5　Honeycomb weaves / 蜂巢组织 ··· 72
4.6　Bedford cord and piqué / 凸条组织 ····································· 76

4.7　Distorted weave effects / 网目组织 ··· 81

Chapter Five　Color and Weave Effects / 色纱组织配合（配色模纹） ············· 87

5.1　Construction of patterns from a given weave and colour repeats
　　根据已知组织图和色纱循环绘作配色模纹 ··· 87

5.2　Construction of weaves from a given pattern and colour repeats
　　已知色纱循环和配色花纹绘作组织图 ··· 105

5.3　Construction of weaves and colour repeats from a given pattern
　　已知配色花纹确定色纱排列和组织图 ··· 106

UNIT II　Compound Structure / 复杂组织 ··· 109

Chapter Six　Backed Weaves / 二重组织 ··· 110

6.1　Warp backed weaves / 经二重组织 ··· 110

6.2　Weft backed weaves / 纬二重组织 ·· 112

Chapter Seven　Multi-ply Fabrics / 多层织物 ··· 118

7.1　Double fabrics / 双层组织 ·· 118

7.2　Stitching double fabrics / 接结双层 ·· 121

7.3　Interchanging double cloth / 双层表里交换 ·· 134

7.4　Tubular cloths / 管状织物 ··· 136

7.5　Double width cloths / 双幅织物 ··· 138

7.6　Multi-layer weaves / 多层组织 ·· 139

Chapter Eight　Pile Fabrics / 绒织物 ·· 152

8.1　Weft pile / 纬起绒织物 ·· 152

8.2　Warp pile / 经起毛织物 ··· 156

8.3　Terry pile / 毛圈织物 ··· 159

Chapter Nine　Gauze and Leno Weaves / 纱罗组织 ···································· 167

9.1　The concept of gauze / 基本概念 ··· 167

9.2　The principles of gauze weave formation / 纱的形成原理 ·························· 167

9.3　Gauze and leno weave examples / 纱罗组织示例 ···································· 170

Chapter Ten　Jacquard Fabrics / 纹织物 ·· 176

10.1　Elements of jacquard shedding / 提花开口原理 ······································ 176

10.2　Preparation for designing the jacquard fabrics / 纹织物设计准备 ··············· 178

10.3　Steps in construction of jacquard design / 纹织物设计步骤 ·············· 180

UNIT Ⅲ　Fabric Design / 织物设计 ·············· 185

Chapter Eleven　Fabric Geometry / 织物的几何结构 ·············· 186
11.1　Fabric geometry / 织物的几何结构 ·············· 186
11.2　Fabric cover and cover factor / 紧度和紧度系数 ·············· 197

Chapter Twelve　Target Design and Materials Selection / 产品定位设计与原料选择 ······ 203
12.1　Target design / 定位设计 ·············· 203
12.2　Material selection / 原料选择 ·············· 205

Chapter Thirteen　Calculation and Selection of Varied Parameters / 工艺参数的选择与计算
·············· 206
13.1　Selection of weaves / 组织选择 ·············· 206
13.2　Yarn linear density / 纱线线密度确定 ·············· 206
13.3　Fabric setting / 织物密度 ·············· 208
13.4　Crimp of yarn in woven fabric / 织物缩率 ·············· 209
13.5　Twist direction and the twill prominence / 捻向与斜纹的明显性 ·············· 211
13.6　Selvedges / 布边设计 ·············· 212

Chapter Fourteen　Example of Fabric Design / 织物设计举例 ·············· 214
14.1　Grey fabric design and calculation / 坯布设计和计算 ·············· 214
14.2　Example / 示例 ·············· 217

Chapter Fifteen　Glossary of Fabrics / 织物生词表 ·············· 219

Appendix Ⅰ　Fabric Design Table / 附录 1　织物设计工艺表 ·············· 234
Appendix Ⅱ　Parameters and Specifications of some Grey Fabrics
　　　　　　　附录 2　部分坯织物规格和技术条件 ·············· 235

References / 参考文献 ·············· 244

UNIT I

General Knowledge and Simple Construction
基本知识和简单组织

Chapter One
General Knowledge on Woven Fabrics
有关机织物的基本知识

1.1 Cloth formation on loom

Woven fabrics are formed by interlacing two mutually perpendicular sets of yarns. The simplest interlacing pattern is illustrated in Fig. 1.1, where eight warp and six weft threads are included.

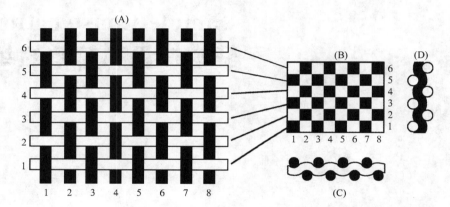

Fig. 1.1 Plan and section of plain weave

A textile designer must understand the influence of the process of cloth formation on fabric structures. In order to interlace warp and weft threads, the loom must carry out the five operations below.

The first three operations are the most important for cloth formations known as the primary operations. The other two operations are essential for continuous fabric production.

(1) Shedding— separating the warp threads into two layers, one of which is lifted and the other is lowered to form the space for the weft insertion, which is called a shed.

(2) Picking— inserting the weft thread through the shed, sometimes by a shuttle.

(3) Beating-up— pushing the newly inserted weft, known as a pick, into the already woven fabric to the point called the fabric fell.

(4) Warp letting-off— delivering the warp to the formation zone at the required rate and at a suitable constant tension by unwinding it from the

UNIT Ⅰ   General Knowledge and Simple Construction

weaver's beam.

(5) Cloth taking-up— moving fabric from the formation zone at the constant rate that ensures the required pick spacing, and winding the fabric onto a cloth roller.

卷取

卷布辊

The schematic diagram of a loom is illustrated in Fig. 1.2, where the principal parts for the five basic motions are shown. The warp yarn 1 from the weaver's beam 12 passes round the back rest 2 and goes through the drop wires 3 of the warp stop-motion to the healds 4, which are intended for separating the warp threads for the purpose of shed formation. It then passes through the reed 5 that holds the threads at uniform spacing and is designed for beating-up the weft thread which is inserted into the triangular warp shed 7 by the shuttle 6. The shed is formed by two warp sheets (layers) and the reed. Temples hold the cloth at the fabric fell 8 to assist in formation of a uniform fabric, which then passes over the breast beam 9, round the take-up roller 10 and onto the cloth roller 11.

织轴/后梁/
停经/片/综丝

筘

边撑
胸梁/卷布辊

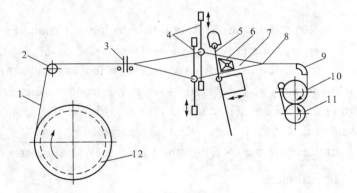

Fig. 1.2   Schematic diagram of the loom

Besides the five basic motions, which perform the above operations the loom is equipped with a number of auxiliary motions, which are intended for automation, and control of the process of weaving, and for increasing the efficiency.

These basic motions ensure the required structure and properties of the fabric by controlling such parameters as the density of weft threads (pick-spacing), density of warp threads, the type of weave, and the crimp of threads.

缩率

The density of weft threads in the fabric is determined by the cloth take-up motion, which controls the rotary speed of the take-up roller.

The number of warp threads per unit of fabric width. i.e. warp density, depends on the count of the reed which is a part of slay motion, weft crimp and the number of threads drawn into one reed-dent.

筘座
筘齿

The influence of warp tension, which is controlled by the let-off motion, on the crimps of threads is well known. The level of warp tension affects the weaving condition, and determines the yarn crimps in the fabric.

The weave type is fully dependent on the type of shedding motion. Tappet shedding motions are used for simple fundamental weaves; dobbies are employed for more complex weaves, such as the derivative and combined weaves. For producing fabrics with large figured patterns, a jacquard shedding mechanism becomes necessary. The figuring capacity of jacquard shedding mechanism may range from hundreds to thousands threads and even more. The number of warp threads with different order of interlacement depends on the number of hooks used in the mechanism that represents its figuring capacity. The number of weft threads with different order of interlacement, i.e. weft repeat of the weave, can be changed by altering the number of perforated cards in the chain which is formed when the last card is joined to the first.

1.2 Fabric analysis

The properties of the fabric are closely linked to fabric parameters including the weave, the arrangement of warp and weft, the raw materials, the density of threads in the fabric, the characteristics of warp and weft threads, the characteristics of fibers and the cloth geometry introduced during weaving, such as yarn crimp. In order to develop fabrics with specific properties, it is imperative to work out all the above fabric's parameters.

The following steps are involved in the fabric analysis process.

1.2.1 Making samples
The sample should properly represent the characteristics of the fabric.
1.2.1.1 Location
The sample should be selected from the middle of the fabric. The distance from the selvage must be more than 5cm, and the distance from the end of the fabric must be more than 1.5 to 3m.
1.2.1.2 Size
The size of the sample varies depending on the characteristics of the fabric. 15cm×15cm are suitable for simple structure and fabrics with small patterns, and 20cm×20cm for fabrics with big patterns.

1.2.2 Identification of face and back of the fabric
We can identify the face or back of the fabric according to the appearance of the fabric. The following points is helpful for efficient identification.
- The face of a fabric has a clear colour or patterns.
- For rib or corded fabrics, the face is usually more dense and smooth.

- For pile fabrics, the face has piles. 绒织物
- For double clothes, the faces have higher density, and use better materials. 双面织物
- For terry fabrics, the faces have denser loops.

1.2.3 Identification of warp and weft

We can identify the warp and weft directions according to the following points.
- The warp has always parallel with selvage.
- The warp may be sized, and the weft is not. 上浆的
- The warp usually has a big density.
- The warp direction may have reed marks. 筘痕
- The warp is more likely to ply yarn. 合股线
- The Z-twist yarns are used for warp, and S-twist for weft if the warp and weft are different in twist directions. Z捻/S捻
- The warp usually has higher twisted yarns.
- The warp usually has good quality yarns.
- For terry fabrics, the warp forms the loops. 圈
- For striped fabrics, the warp is in parallel with the stripes. 条
- Warp can be easily arranged with different kinds of yarns.

1.2.4 Density measurement

The density of fabric is very important, as it directly affects the fabric's appearance, handle, thickness, strength and warmth retention. There are two different methods for measuring the fabric density. 手感

1.2.4.1 Direct measurement

The fabric sample is placed on a flat surface, making sure it is not under tension or distorted. A piece glass is placed on top and the fabric is viewed through the lens. A magnified image of threads occupying the length and width of the piece glass square can be counted with the aid of a dissecting needle to pinpoint the individual threads in warp and weft.

If the piece glass is $2.5 cm^2$, the threads in one direction of woven fabric will be determined by:

Thread density = threads counted /2.5 = threads per cm

1.2.4.2 Indirect measurement

This is carried out using an optical device known as a taper line grating. This is a flat sheet of glass with a large number of straight lines engraved on it in a tapered fashion so that their density increases from left to right when this is placed on top of a simple woven construction the threads interfere with the line grating and an optical pattern is produced.

1.2.5 Crimp measurement

Crimp refers to the amount of bending that is done by a thread as it interlaces with the threads that are lying in the opposite direction of the fabric.

Fig. 1.3 shows the difference between the length of yarn (l_y) taken from a length of fabric (l_f), so

$$Crimp\ (c) = \frac{l_y - l_f}{l_f}[1].$$

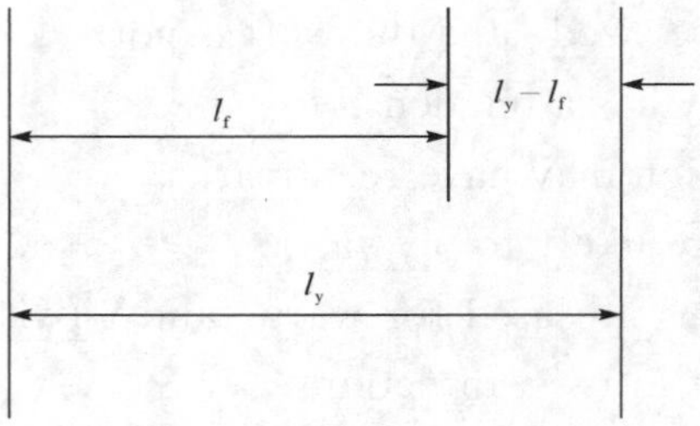

Fig. 1.3 Crimp

Often it is generally considered most convenient and preferable to use percentage values:

$$Crimp\ (c) = \frac{l_y - l_f}{l_f} \times 100/\%.$$

The use of calculations using the crimp formula is essential in determining the amount of yarn that is required for a particular circumstance or in assessing how much fabric can be produced from a known length of yarn. 缩率测试仪 Fig. 1.4 shows a <u>crimp tester</u>.

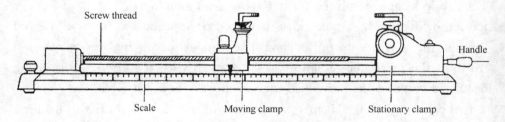

Fig. 1.4 Crimp tester

线密度 ### 1.2.6 Linear density (tex) measurement

It is common to specify thread thickness as thread linear density in tex. The thread linear density indicates the weight in grams of 1km of thread. The formula is:

---

[1] 此公式和中文教材中的缩率有区别——编者注.

$$Linear\ density\ (\text{tex}) = \frac{1\,000\,G}{L},$$

here: G — the weight of the thread at the nominal regaing(g);
L — the length of the thread(m).

1.2.7 Fiber identification

The approach for fiber identification depends on whether a sample consists of one type of fiber or a blend of fibers. In the case of blends the quantitative analysis of the fibers may also be necessary. Fiber identification may be carried out using the following methods.

(1) Microscopically examination of the longitudinal and cross-sectional views of the fiber.
(2) Burning test.
(3) The use of solvents.
(4) Other chemical tests.
(5) Staining.
(6) Melting point determination.
(7) Fiber density.

It should be remembered that sometimes no single method can give a completely reliable indication of the identity of the fiber and that confirmatory tests are often advisable.

In order to determine whether more than one type of fiber is present, microscopical examination of the fiber is convenient.

1.2.8 Fabric weight

Fabric weight is economically important. It can be measured in two ways.

1.2.8.1 Weighting measurement

As absorbed moisture affects both mass and dimensions, it is important to precondition samples and carry out measurements in a <u>standard atmosphere</u>.   标准温湿度

Mass per unit area is expressed in grams per square meter (g/m²), but it is not necessary to measure a square meter of fabric. A relatively small <u>specimen</u> is cut out, usually 10cm × 10cm, with the aid of a template. The   标样
specimen is then weighed accurately.

$$Mass\ per\ unit\ area\ (\text{g/m}^2) = specimen\ mass(\text{g}) \times 100$$

1.2.8.2 Calculating measurement

Fabric weight is the weight of yarn per square meter in a woven fabric, which is the sum of the weight of the warp and the weight of the weft, so:

$$W\ (\text{g/m}^2) = \frac{10P_1 \times Tex_1}{(1-a_1) \times 1\,000} + \frac{10P_2 \times Tex_2}{(1-a_2) \times 1\,000},$$

here: P_1 — warp density (ends/10cm);
P_2 — weft density(picks/10cm);
a_1 — warp take-up(%);
a_2 — weft take-up(%);
Tex_1 — warp linear density;
Tex_2 — weft linear density.

1.2.9 Fabric structure and colour arrangement

It is necessary to know the method of fabric construction, e.g., a twill weave or a sateen weave. This is determined by viewing the fabric using a piece glass or a low magnification microscope.

1.3 Fabric representation

1.3.1 Methods of weave representation

The interlacing pattern of the warp and weft is known as the weave. Two kinds of interlacement are possible. The first kind—warp over weft—is called warp overlap and the second kind of interlacement—weft over warp—is called weft overlap.

The interlacement is achieved by movement of the warp threads. The warp thread must be lifted to obtain a warp overlap; in this case the weft thread is inserted under the warp. When the warp thread is lowered, the weft thread is inserted above this warp thread and a weft overlap is obtained.

In order to identify the weave of a fabric, it is necessary to look at the face side of a fabric through a magnifying glass. The fabric, shown in Fig. 1.1 (A), contains eight warp threads and six weft threads. At the point of intersection of warp thread 1 and weft thread 1, the warp thread passes over weft thread 1. This is a warp overlap. Examining the intersection of warp thread 2 and weft thread 1, one can see that the weft thread 1 is placed over the warp thread 2. This is a weft overlap. A weave is characterized combination of warp and weft overlaps arranged in certain orders. However, drawing such diagrams is a time consuming operation and, therefore, this method is not used by textile designers.

A practical method of weave representation called canvas method is widely used internationally. In this method, a squared paper, design-paper is employed. On this design-paper, shown in Fig. 1.5, each vertical space represents a warp thread, and each horizontal space a weft thread. Each square, therefore indicates an intersection of warp and weft threads. To show the warp overlap, a square is filled in or shaded. The blank square indicates that the weft thread is placed over the warp, i. e. a weft overlap. Several types of marks may be used to indicate the warp overlap. In some cases digit 1 is used to indicate the warp overlap, and digit 0 to indicate the weft overlap, and the weave diagram becomes a matrix which is a convenient form for the computer. For interlacement, the threads must cross each other, passing over or under the threads of another system. Therefore, in a full weave repeat every vertical space and every horizontal space must have at least one mark and at least one blank, otherwise the threads do not interlace but form <u>loose floats</u> on the face or back side of the fabric. A weave diagram is shown in Fig. 1.1 (B), corresponding to the weave illustration in Fig 1.1 (A). Warp thread 1 passes over the first weft thread, under the second, over the third, etc.; warp thread 2, under the first, over the second, under the third, etc. Warp thread 3 repeats the pattern of interlacement of the first warp thread; the fourth warp thread repeats that of the second thread. Therefore, the <u>warp repeat</u> of this weave equals 2. Examining the order of interlacement of weft threads, one will find that the <u>weft repeat</u> of this weave also equals 2.

浮线

经循环
纬循环

The <u>plan</u> of weave in Fig. 1.1 (B) does not indicate the configuration of the threads within the fabric. When it is necessary, the weave may be supplemented by cross-sectional views either in warp or weft direction. Cross-section along the weft direction at weft thread 1 is shown in Fig. 1.1 (C). Black circles represent the warp threads and configuration of weft thread 1 is shown between the warp threads. Cross-section along the warp direction at warp thread 8 is shown in Fig. 1.1 (D). Black circles represent the weft threads and warp thread 8 passes between weft threads.

平面图

Fig. 1.5 shows the different marks together with what they mean.

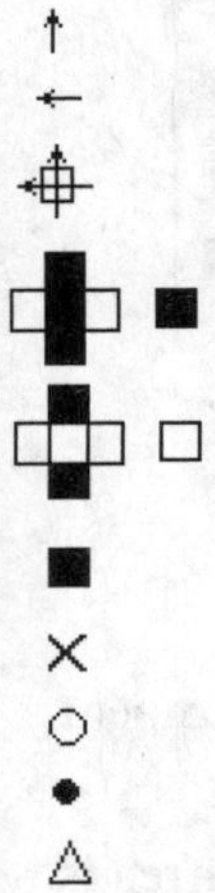

Each vertical space represents a warp end.
Each horizontal space represents a weft pick.
Each square indicates an <u>intersection point</u> of one end and one pick.   组织点
Warp floats = Warp over weft.
Only warp floats or lifts are indicated.
Weft floats = Warp under weft.
Blanks represent weft floats.

In some structures several different marks are used simultaneously.
It is important to give a clear indication of the key to a diagram.
All marks = Warp up.

Fig. 1.5 Representation of marks

组织循环与飞数/组织循环

经纱循环

纬纱循环

1.3.2 Weave repeat and shift

The repeat of a weave is a complete unit of the weave. And it is also called a weave repeat. All weaves repeat on a certain number of warp threads and weft threads. It is sufficient to show only one repeat of a weave on the design-paper. Weave repeat is used as a basis to construct the woven fabric of required size. It is defined by warp repeat and weft repeat. The warp repeat is a minimal number of warp threads after which the sequence of warp threads with a different interlacing repeats. The weft repeat may be defined analogically. In some weaves the warp repeat is equal to the weft repeat whereas, in others the warp repeat differs from the weft repeat. The weave of fabric is determined by a certain arrangement of overlaps. In the canvas method of weave representation the weave appears in an abstract mathematical form as a combination of painted and blank squares. The arrangement of painted squares characterizes the weave. On the design-paper the warp repeat is equal to the number of vertical spaces or squares in the repeat of design, and the weft repeat, to the number of horizontal spaces. When the weave is presented on design-paper (Fig. 1.6), the warp repeat can be found by comparing the vertical spaces. There is a certain arrangement of painted squares in each vertical space. In the first vertical space two painted squares are located in the horizontal spaces 1 and 6; in the second, spaces 4 and 9; in the third, spaces 2 and 7, and so on. In the sixth vertical space the arrangement of painted squares is the same as in the first one, in the seventh- as in the second, and then in the tenth-as in the fifth. Thus, the weave repeats after five vertical spaces, and warp repeat of this weave is 5. Denoting the warp repeat by R_0, we have $R_0 = 5$.

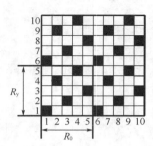

Fig. 1.6　Weave repeat

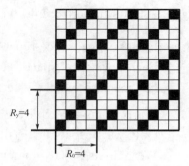

Fig. 1.7　Twill weave repeat

Similarly, the weft repeat can be found. Let R_y represent the weft repeat, and we have $R_y = 5$.

The weave represented in Fig. 1.7 contains 12 warp and 12 weft threads. Looking at the weave one can find that the warp repeat, as well as the weft

UNIT I General Knowledge and Simple Construction

repeat, equals 4. It is possible to pick out the weave repeat in different places, as it is shown by the thick line. The weave repeat appears different owing to changing the starting point, but it has no influence on the appearance of the fabric. Nevertheless, for uniform weave representation it is advisable to start constructing the weave from the point of intersection of the first warp thread and the first weft thread.

One repeat contains the smallest number of intersections between warp and weft (or ends and picks) which, when repeated in either direction, give the weave of the whole fabric(see Fig. 1.8)

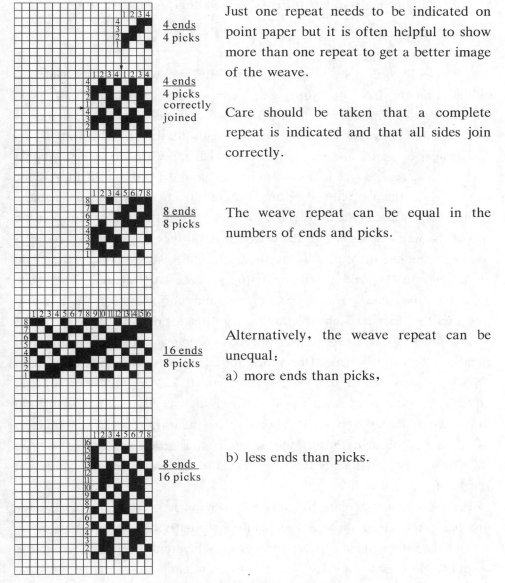

Just one repeat needs to be indicated on point paper but it is often helpful to show more than one repeat to get a better image of the weave.

Care should be taken that a complete repeat is indicated and that all sides join correctly.

The weave repeat can be equal in the numbers of ends and picks.

Alternatively, the weave repeat can be unequal:
a) more ends than picks,

b) less ends than picks.

Fig. 1.8 Repeat of various weaves

飞数
对应点

Full characteristic of a weave includes not only the repeat, but also a shift or move. The shift is the distance from a painted square on one horizontal space or vertical space to the corresponding painted square on the next horizontal space or vertical space. It is common to use the number of square as a unit of length in determining the shift. Due to this, the shift can be 0, 1, 2, 3, etc. One characteristic feature of all the fundamental weaves is that they all have a constant shift. For the plain and twill weaves $S_y = 1$, where S_y is the shift along the horizontal space. Crepe weave has variable shifts. Shift can be positive and negative, depending on the direction of counting. Counting from left to right or upward gives positive shift.

穿综图

1.3.3 Drafts
1.3.3.1 Draft types

综框
综
综眼

The weave is formed by interlacing the warp and weft threads, which is achieved on the loom by raising and lowering the heald frames, or heald shafts, a set of which is called harness. Each shaft contains a big number of healds, through eyes of which the warp threads are drawn. The draft shows the number of shafts and the manner in which the warp threads are drawn into the shafts. The type of weave to be produced depends on the draft. The drawing-in of the warp threads is done before starting the weaving. The basic data for constructing the draft is the weave. The draft is usually shown at the top of the weave diagram. In the canvas method the squared design paper is used to represent the draft, where the horizontal spaces represent the heald frames, or shafts, and the vertical spaces, the warp threads. The circle placed in the square where the vertical and horizontal spaces intersect indicates the shaft through which the warp thread is drawn.

穿经
组织图

The draft is characterized by its repeat which is equal to the minimal number of vertical spaces after which the order of drawing-in the threads repeats. The repeat of draft in Fig. 1.9 equals 5. The repeat of draft is equal, as a rule, to the warp repeat. There is all exception in the case of skip draft, when the repeat of draft is greater than the warp repeat 2 or 3 times to avoid having too many healds on one shaft, thus reducing friction between the threads. But in all cases the warp repeat of the weave cannot exceed the repeat of draft.

综片数

The other characteristic of draft is the number of shafts necessary for producing the given weave. The number of shafts can be equal to or be greater than the number of warp threads with different interlacing within the warp repeat. Each thread with a different interlacement requires an individual shaft. Therefore, the minimal number of shafts is equal to the number of threads in the warp repeat having different order of interlacement.

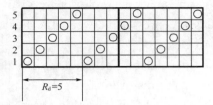

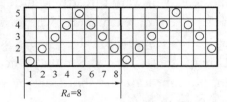

Fig. 1.9 Straight draft   Fig. 1.10 Point draft

The number of vertical spaces of the draft should be the same as the number of vertical spaces of the weave diagram. The leftmost vertical space of the draft is taken as the first one and the numbering is done from left to right (Fig.1.10). While numbering the shafts, the front shaft which is the nearest to the reed is usually taken as the first. Thus, the first horizontal space in the draft is the nearest to the weave diagram.

The various drafts can be classified as follows.

- ◆ Straight   ◆ Divided
- ◆ Skip and sateen   ◆ Grouped
- ◆ Pointed   ◆ Curved
- ◆ Broken   ◆ Combined

顺穿/分区穿
飞穿/分组穿
山形穿/照图穿
人字穿/联合穿

Straight draft The straight draft is the simplest type of draft. It forms the basis for many other drafts. Straight draft is the most common and can be used with any number of shafts. Each successive thread is drawn on successive shaft, the first thread on the first shaft, the second thread on the second shaft, and so on (see Fig.1.9). The last thread of the warp repeat is drawn on the last shaft. Thus, the number of shafts equals the warp repeat, and the repeat of draft equals the warp repeat, i.e. $R_d = R_0 = 5$. Oral instructions for straight draft are sufficient for drawers and operators.

逐一顺次

Skip draft This is used in weaving the fabrics with a high density of warp threads. It makes it possible to use a number of shafts two or more times greater than the warp repeat and than the minimal necessary number of shafts for this weave. The density of healds on each shaft decreases, and friction thread against thread, and thread against heald reduces.

The skip draft on six shafts (Fig. 1.11) is used for the plain weave with warp repeat $R_0 = 2$. For constructing the draft, the shafts are divided into two groups equal to the warp repeat. The odd threads 1, 3, 5 are drawn on the shafts of group I. Each successive thread is drawn on successive shaft of this group. The even threads 2, 4, 6 are drawn in group II of shafts in the similar way.

It is known that the minimal number of shafts for the plain weave is 2, and if the density of warp threads in fabric per cm is, for example, 66, it gives 33 healds per cm on each shaft. The threads are pressed between the healds, and due to great friction it is impossible to weave.

Using 6 shafts, one gets only 11 healds per cm on each shafts. Besides, with the skip draft the shafts are moved in groups: the shafts 1, 2, 3 are lifted forming the first shed, and the shafts 4, 5, 6, forming the second shed, and so on. Only two tappets in the shedding motion can be applied in this case.

The sateen draft is used instead of skip one with the same purpose when the warp repeat is greater than 5. In sateen draft the number of shafts equals the warp repeat. The uniform arrangement of healds is achieved by distributing them in the harness in the manner of warp overlaps of sateen weave.

Pointed draft   This draft is used in weaving the fabrics with a symmetrical design, when the straight draft cannot be applied because of a large warp repeat. The pointed draft can be considered as a combination of straight drafts, constructed first in one direction and then in the other. The change of direction takes place on the first and last shafts of the draft, which are the points of reversal. These shafts contain only one heald each within the draft repeat (Fig. 1.12). Each other shaft contains two healds. The number of shafts used is one more than half the warp repeat of the weave: $N_s = R_0/2 + 1$. A variety of derived twills can be produced by using the pointed draft, such as waved twills, and also the diamond designs, which are symmetrical about their vertical and horizontal axes.

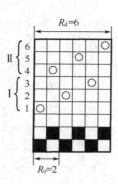
Fig. 1.11 Skip draft

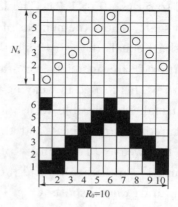
Fig. 1.12 Waved twill and point draft

Broken draft   This draft can be considered as a modified pointed draft. Again, this is a combination of straight drafts with different directions of constructing, but the direction is reversed not on the last or the first shaft. When the direction is reversed, the first thread of the next group is started higher or lower than the last thread of the preceding group. This small modification changes considerably the design by breaking the axis of symmetry. The order of interlacing of the last thread of the first group is opposite to that of the first thread of the preceding group. The warp thread 9 (Fig. 1.13), the first thread of the second group, has the order of interlacing opposite to that of thread 8, i. e. the warp overlaps of thread 9 are at the points of intersection with wefts 2, 3, 6, 7, and those of thread 8 with wefts 1, 4, 5, 8. The broken draft is applied for producing some <u>herringbone twills</u>, <u>diaper designs</u>, and some other weaves.

海力蒙/菱形花纹

Divided draft　　This draft is employed for derived weaves, <u>double-warp weaves</u>, <u>two-ply weaves</u>, <u>pile weaves</u>, and other weaves involve two sets of warp yarns.

经二重/双层/绒织物

The shafts are divided into two or more groups. A suitable type of draft is chosen for each group. The divided draft, shown in Fig. 1.14 is employed for a double-warp fabric. There are two systems of warp threads: the face and back ones. The odd face threads are drawn on the shafts of 1, 2, 3, 4. The even back threads are drawn on the shafts of 5, 6, 7, 8, 9, 10, 11, 12. It is common to place the shafts with more frequently interweaving threads at the front. The draft for each group is chosen a straight one.

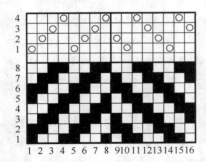

Fig. 1.13　Waved twill and broken draft

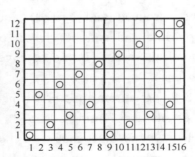

Fig. 1.14　Divided draft

Grouped draft　　This draft is employed for production of <u>check and stripe</u> designs, in which the stripes have different weaves or their combinations. The draft in Fig. 1.15 is used for producing the fabric with two different stripes containing 15 and 12 threads, respectively.

条格

The warp repeat of the first stripe equals 3, and the second equals 4. The weave of the first stripe requires 3 shafts, and that of the second stripe requires 4 shafts. All the threads of the first stripe are drawn on three first shafts with straight draft, and the threads of the second stripe are drawn on

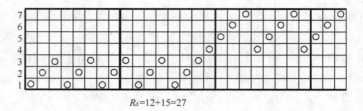

$R_d=12+15=27$

Fig. 1.15 Grouped draft

shafts 4, 5, 6, 7 of the second group. The repeat of draft is 27. The repeat of draft is determined by the number of stripes and the number of threads in each stripe. The number of shafts in the draft depends on the number of stripes and the warp repeat of weave of each stripe.

Curved drafts These drafts are applied for <u>fancy weaves</u> having a large warp repeat with the purpose of reducing the number of shafts. Note, that the minimal number of shafts equals the number of threads in warp repeat with different order of interlacing. The drawing-in is done applying the rule: all warp threads which work alike are drawn on the same shaft. Curved drafts are irregular and cannot be classified. Oral instructions cannot be given to the <u>drawers</u>, as well as to the <u>weavers</u>. They should be supplied with drawings of the curved draft.

Combined draft Various methods of drawing-in can be combined in one draft for producing a certain type of fabric. Two or more drafts described above can be applied simultaneously, for example, straight and skip or sateen, grouped and curved, and so on. Combined draft is the most complicated and can be chosen only if there are some technological or economical reasons. It can be done properly by the designer having a great experience.

1.3.3.2 Requirements to drawing-in

The possibility of using a straight draft should be studied first. Then the simplest type of draft should be chosen, suitable for the given weave.

The number of shafts should be as small as possible, but the density of healds is to be calculated and should not exceed the standard value.

The distribution of threads on different shafts should be as uniform as possible. In some cases additional shafts can be added to reduce the density of healds.

It is advisable to use the front shafts with the minimum <u>height of lifting</u> for the threads with the biggest number of intersections in the weave repeat and for a weaker system of warp threads.

1.3.4 <u>Weaving plan</u>

The weaving plan consists of three elements placed in a certain order: (1) the

weave, (2) the draft, and (3) lifting plan (Fig.1.16(A)).

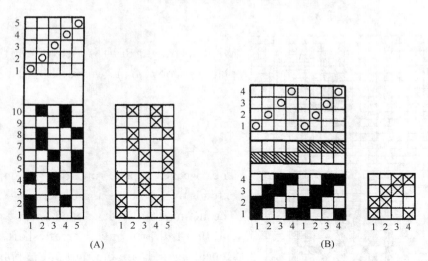

(A) (B)

Fig. 1. 16 Weaving plan with straight draft

These <u>elements</u> are closely interrelated to one another. If any two elements of the weaving plan are known, the third element can be constructed. The weave is determined entirely by the draft and the order of lifting the heald frames, or shafts. Changing either the draft or the lifting plan results in a new weave.

In some cases an additional element of weaving plan can be shown. This is a drawing of the warp threads through the reed, or <u>dents</u>. The drawing through the reed indicates the number of warp threads per reed dent. This element of weaving plan is positioned between the weave repeat and the draft [see Fig. 1.16(B)].

1.3.5 Lifting plan

In order to produce the required weave the designer has to provide a lifting plan for the purpose of controlling the lifting and lowering of the shafts. In <u>dobby shedding</u>, the plan is used either for pegging a set of lags or cutting a paper card. Lifting plans are indicated on the right-hand side of the weave diagrams in Fig.1.17.

1.3.6 Relations between weave, draft, and lifting

The three elements of a weaving plan are dependent on one another. Any one element of the weaving plan can be constructed if two others are given.

1.3.6.1 <u>Construction</u> of a lifting plan from the given weave and draft

This has been described in section 1.3.5.

1.3.6.2 Construction of draft from the given lifting plan and weave

Fig. 1. 18 shows that the lifting plan and the weave are given, and the problem is to construct the draft.

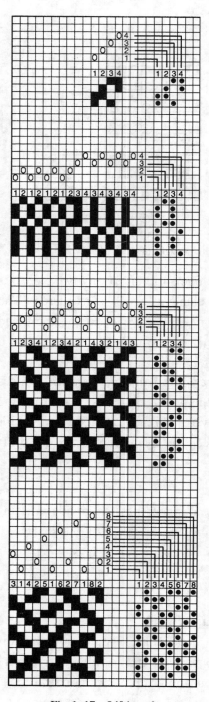

When a straight draft is used, the lifting plan is a copy of the weave.

The weave repeat on this example extends over 16 ends and 8 picks.
The lifting plan is developed from each end that is drawn on a separate shaft.
Repeating ends are omitted in the plan.
Number of shafts required: 4.

Weave repeat: 16 ends and 16 picks.
Number of shafts required: 4.

Weave repeat: 12 ends and 12 picks.
Number of shafts required: 8.

Fig. 1.17 Lifting plan

UNIT I General Knowledge and Simple Construction

The number of vertical spaces in Fig. 1.18 (C) corresponds to the number of horizontal spaces in Fig. 1.18 (B). It means that the first vertical space in Fig. 1.18 (C) controls the first shaft, the second vertical space controls the second shaft, and so on. Then the draft is constructed by comparing the arrangement of the lifting plan in Fig. 1.18 (C) and weave in Fig. 1.18 (A). The movement of the warp thread 1 is same as the first shaft. So, the warp thread 1 must be drawn in first shaft. Similarly, the movement of the warp thread 2 is the same as the third shaft, so it must be drawn in third shaft. And so on.

1.3.6.3 Construction of weave from the given draft and lifting plan

Fig. 1.19 shows that the lifting plan and draft are given, and the problem is to construct the weave. The number of vertical spaces in Fig. 1.19 (C) corresponds to the number of horizontal spaces in Fig. 1.19 (B). The weave can be constructed by comparing the lifting plan in Fig. 1.19 (C) and draft in Fig. 1.19 (B). The warp thread 1 is drawn in shaft 1, so the movement of the warp thread 1 should be the same as the shaft 1. The warp thread 2 is drawn in shaft 3, so the movement of the warp thread 2 should be the same as the shaft 3. We can construct the warp threads 3, 4 in similar way.

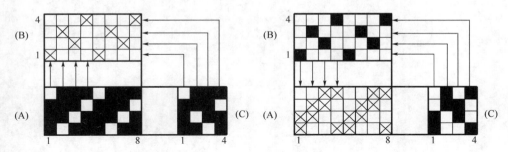

Fig. 1.18 Construction of draft   Fig. 1.19 Construction of weave

HOMEWORKS

1. Explain the five motions briefly.
2. Draw the schematic diagram of weaving and describe it.
3. Analyze one or two fabric samples, and get all the parameters you could.
4. Explain the concepts: repeat, warp repeat, weft repeat, weave, shift.
5. Construct lifting plan from given weave and draft (See Fig. 1.20).
6. Construct weave from given draft and lifting plan (See Fig. 1.21).

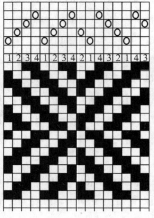

Fig. 1.20

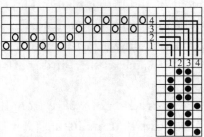

Fig. 1.21

Chapter Two
Elementary Weaves (Fundamental Weaves)
三原组织

2.1 General characteristics

基本特征

The fundamental weaves possess the following characteristics.

(1) The shift of the weaves is <u>constant</u>.

常数

(2) The fundamental weaves are such weaves where every warp and weft thread within the repeat interlaces or is <u>interlaced</u> only by one thread of the opposite system. It means that, in fundamental weaves any weft or warp threads must have only one <u>warp or weft overlapping</u> within the repeat. Consequently, the number of warp threads in a repeat must be equal to the number of weft threads, i.e. $R_0 = R_y = R$, where R is general weave repeat; R_0 is repeat of warp; R_y is repeat of weft.

叠盖

经/纬组织点

There are three types of fundamental weaves: plain, twill and <u>sateen</u>. Each type of fundamental weave is determined by two parameters, and each weave has its own values of parameters. Such parameters are: the repeat R of weave and the shift S of interlacement, or move.

缎纹

The shifts can be counted in both vertical and horizontal directions.

(1) <u>Vertical shift</u>(S_0) — the shift of two warp threads with respect to each other. This shift is often called the warp shift where the count is made in vertical direction.

经向飞数

(2) <u>Horizontal shift</u>(S_y) — the shift of two weft threads with respect to each other. This shift is often called the weft shift where the count is made in the horizontal direction.

纬向飞数

The shift can be <u>negative</u> or <u>positive</u>. The vertical shift is positive, when the count is made upwards. The horizontal shift is positive, when the count is made from left to right, and negative when the count is made from right to left.

正数/负数

2.2 Plain weave

平纹组织

This is the simplest weave. In this weave the threads interlace in alternate order, i.e. the first warp thread crosses over the first weft thread (Fig. 2.1) and passes under the second weft thread; and the second warp thread passes under the first weft thread and crosses over the second one, and so on. The plain weave ensures the simplest form of interlacement of two sets of yarns in such a way that the yarns interlace each other at right angles. The plain weave produces equal number of warp and weft floats in a weave unit.

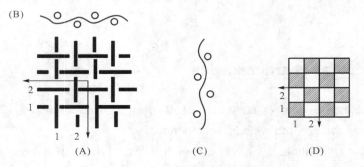

Fig. 2.1 Plain weave and section diagram

In Fig. 2.1, (A) is indicates the interlacement, (B) is cross section diagram along the weft direction, (C) is cross section diagram along the warp direction, and (D) is weave diagram.

A plain weave fabric can be either balanced or unbalanced. In balanced fabrics the warp and weft counts are similar, and the warp and weft densities of the fabric are also similar. The warp and weft crimps are usually the same. Plain weave fabrics are widely used, much more than fabrics of any other weaves.

In the plain weave the values of parameters are the simplest: repeat, $R = 2$; shift, $S = 1$ (see Fig. 2.1(A)). There are only two threads with opposite interlacement within the repeat.

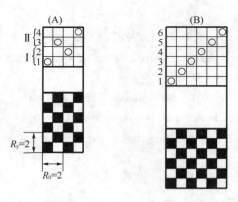

Fig. 2.2 Plain weave with skip and straight drafts

Two heald shafts are sufficient to produce a plain weave fabric. When the number of ends per centimeter is large, 4 or 6 heald shafts (Fig. 2.2) are used in the skip draft.

Usually, the tappet shedding mechanism is used for producing plain weave fabrics. 踏盘开口

Because the plain weave has the maximum number of interlacement possible, the fabric will be firmer and stronger than a fabric otherwise identical but made in a twill weave.

A variety of decorative effects can be produced in a plain-weave cloth. For instance, stripes or checks.

The properties of a fabrics are governed not only by the weave used but also by the weight of the cloth, how close or open the fabric is, the ratio of the number of ends to picks per centimeter, the type of yarn used, the fiber content, and also the finish. 密/稀

Some fabrics made from the plain weave are listed below(Table 2.1).

Table 2.1 Fabrics made from plain weave

Group 1	Group 2	Group 3
batiste/ 细布	A fgalaine/ 平纹毛呢	chiffon/ 薄绸
buckram/ 硬衬布	crepe/ 绉	grosgrain/ 罗缎
cambric/ 细纺	delaine/ 薄花呢	jasper/ 芝麻绸
lawn/ 上等细布	domet(domett)/ 双面厚绒布	organza/ 透明纱
organdie(organdy)/ 暗翼纱	flannel/ 法兰绒	poult/ 波纹绸
poplin(popeline)/ 府绸	panama/ 巴拿马薄呢	shantung/ 山东绸
Voile/ 巴里纱	thornproof tweed/ 防刺粗呢	taffeta/ 塔夫绸
	tropical suiting/ 薄形精纺呢	

The fabrics in Group 1 were commonly produced from cotton or linen, those in Group 2 from wool, and those in Group 3 from silk. Nowadays some of those fabric may also be made from manufactured fibers.

2.3 Twill weaves
斜纹组织

Twill weaves produces diagonal lines on the cloth. These weaves are employed for the purpose of ornamentation, and to make cloth heavier, of closer setting, and better drapability can be produced with the same yarns in plain weave. The parameters of twill weaves, i.e. repeat (R) and shift (S) are the following: 装饰 悬垂性

$$R \geqslant 3; S_0 = S_y = \pm 1$$

A twill cannot be constructed with two threads, but with any number which is more than two. The simplest twill contains three ends and three picks. In the repeat of elementary twill the number of picks equals the number of ends.

Shown in Fig. 2.3 is the construction of twill weaves with repeat $R = 4$, but with different signs of shift: $S_y = +1$ and $S_y = -1$. The direction of twill depends on the sign of shift.

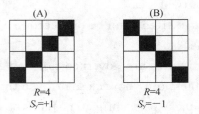

Fig. 2.3　A-right-hand twill,　B-left-hand twill

正面/反面　　Twill lines are formed on both sides of cloth. The direction of diagonal lines on the <u>face side</u> of cloth is opposite to that on the <u>other side</u> coinciding respectively with the weft and warp floats. Thus, if warp floats predominate on one side of the cloth, weft floats will predominate in the same proportion on the other side.

Twill weaves are expressed in the form of a fraction. The numerator of the fraction is equal to the number of warp floats and the denominator is equal to the number of weft floats within the repeat. The sum of the numerator and denominator of this fraction is the repeat of the twill.

Let's take twill $\frac{1}{3}$ for instance. The number of warp floats within the repeat is 1, that of weft floats is 3, and the repeat is 4 (Fig. 2.3).

纬面右斜纹　　When the shift is positive (Fig. 2.3(A)), the single floats form a diagonal which runs from left to right. This is the so-called <u>weft-face right-hand</u> twill.

左斜纹　　When the shift is negative ($S_y = -1$) the diagonal runs from right to left (Fig. 2.3(B)). This is a <u>left-hand twill</u>.

Now we consider twill $\frac{3}{1}$. The repeat of this weave is also 4. On each thread within the repeat there are 3 warp floats and 1 weft float. This is a

经面效果　　warp-face twill. This weave has a <u>warp effect</u> in contrast to twill $\frac{1}{3}$ which has a weft effect, since the proportion of warp floats to weft floats within the repeat is 1 : 3. While designing the fabric with a warp effect, it is preferable to select the density of warp threads higher than that of weft threads, and vice versa.

It is common practice to use straight draft for producing twill fabrics.

组织图　　The following steps are commonly used for drawing a <u>weave diagram</u> (see Fig. 2.4)

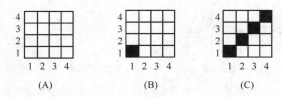

(A) Calculate the repeat R, draw the outline, and give the number.
(B) Draw the first end according to the fraction given.
(C) Draw the other ends according to the shift.

Fig. 2.4　Drawing $\frac{1}{3}$ ↗ twill weave diagram

The prominence of the twill lines in a simple twill weave depends to some extent on the length over which the threads "float" across each other: a $\frac{2}{1}$ twill has a shorter up float than a $\frac{4}{1}$ twill. There is, of course, a limit to the length of float that can be used, since fabrics with very long floats may be unstable and liable to snag. The use of warp and weft yarns differing in colour, diameter, or twist way also enhances the twill effect.　　明显度,效果

The angle at which the twill line runs is influenced by the ratio of end to pick densities and stepping number (shifting number) used in the weave. A stepping number of one in a fabric of square construction will produce a twill line at a angle of 45°. If the ends per centimeter are greater than the picks per centimeter, the twill line will be steeper and will run more closely to the warp direction.

Twill weaves are often used simply to create surface interest. It is possible for the fabric designer to use attractive or expensive yarns in the warp, which will show, and incorporate the less attractive or inexpensive yarns in the weft, which may be scarcely visible on the face of the cloth.

Some fabrics made from twill weaves are listed bellow(Table 2.2).

Table 2.2　Fabrics made from twill weave

Group 1	Group 2	Group 3
denim/ 劳动布	flannel/ 法兰绒	foulard/ 薄软绸
jean/ 三页细斜纹布	gabardine/ 华达呢	surah/ 斜纹软绸
regattas/ 里格特条子布	whipcord/ 贡呢、马裤呢	
regina/ 里津纳细斜纹布	glen check/ 小方格花纹布	
Silesia/ 西里西亚里子布(纬面三页斜纹)		
Drill/ 卡其(斜纹布)		

2.4 Sateen/Satin weaves

In pure sateen weaves the surface of the cloth consists almost entirely of weft floats, as in the repeat of weave each thread of weft passes over all and under one thread of warp.

In addition, the interlacing points are arranged so as to allow the floating threads to slip and cover the binding point of one thread by the float of another, which results in the production of fabric with a maximum degree of smoothness and luster and without any prominent weave features.

The parameters of sateen weaves are: $R \geqslant 5$; $1 < S < (R-1)$.

Besides, for the construction of the regular sateen the shift (S) and repeat (R) must be expressed by prime numbers, i.e. they must not have a common divisor but unity.

The sateen weave is denoted by a fraction. The numerator of this fraction is equal to the repeat of weave. The denominator is equal to the shift (S_y) of overlaps.

Figure 2.5 (A), (B) represents sateen $\frac{5}{3}$ and $\frac{7}{3}$ called 5 ends sateen and 7 ends sateen.

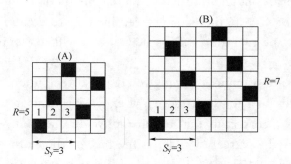

Fig. 2.5 A — sateen $\frac{5}{3}$, B — sateen $\frac{7}{3}$

The warp-faced fabric is called satin, which can be constructed using the vertical shift (S_0) (Fig. 2.6). Satin has the warp effect, and the density of warp is much higher than the density of weft, i.e. P_0 is much higher than P_y. From the weave point of viewer, the back of a satin weave is a sateen weave. It is common practice to weave a sateen to get a satin fabric as less as warp ends need to be raised in weaving a sateen.

The sateen has a weft effect, and $P_y > P_0$.

The following steps are commonly used for drawing the satin/sateen weave diagram.

UNIT Ⅰ General Knowledge and Simple Construction

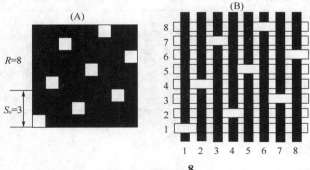

Fig. 2.6 Satin $\frac{8}{3}$

(1) Draw the outline according to the repeat, and give the number of the threads.

(2) Draw the first end or pick. (Satin for the first end, sateen for the first pick).

(3) Draw the other ends or picks according to the shift.

Satin fabrics are available in a range of weights suitable for lightweight <u>linings</u>, <u>lingerie</u>, <u>slippers</u>, foundation <u>garments</u>, heavy dress fabrics suitable for evening wear, and <u>bridal gowns</u>. Generally, the heavier satins are made with an 8-end repeat, and the threads per centimeter and cloth cover are greatly increased.

衬里、女内衣/睡衣/服装/新娘服

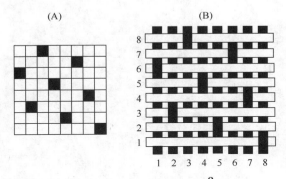

Fig. 2.7 Back side of satin $\frac{8}{3}$

The smooth surface of satin lends itself to <u>printing</u>, machine <u>embroidery</u>, and <u>embossing</u> to produce lustrous effects. Sateens are frequently made from cotton, and a part from being printed and being suitable for dress fabrics, the weave is used in such fabrics as italian for lining and closely woven <u>down proof</u> sateens for <u>eiderdowns</u>.

印花/机绣浮雕印花

防羽绒逸出/鸭绒服

Some fabrics made from satin and sateen weaves are listed below (Table 2.3).

Table 2.3 Fabrics made from satin and sateen weaves

Group 1	Group 2	Group 3
italian/ 意大利缎	doeskin cloth/ 驼丝锦	duchesse satin/ 丝硬缎
satin drill/ 直贡	satin-back gabardine/ 缎背华达呢	satin-back crepe/ 缎背绉
venetian/ 威尼斯八枚经面缎		satin marocain/ 波纹缎

HOMEWORKS

1. Drawing the following weave diagrams, and indicates the warp face twill or weft face twill.

$$\frac{1}{3}\nearrow, \frac{4}{1}\nwarrow, \frac{2}{1}\nearrow, \frac{3}{1}\nearrow$$

2. Listing 5 fabrics made from plain weave and 5 fabrics made from twill weave.

3. Drawing the following weave diagrams:

$\frac{5}{3}$ sateen, $\frac{8}{3}$ sateen, $\frac{5}{3}$ satin, $\frac{11}{3}$ sateen.

Chapter Three
Derivatives of Elementary Weaves
变化组织

These weaves are constructed by means of developing elementary weaves. They are derived by changing the floats, number of shift, direction of diagonal lines, from plain, twill, and sateen/satin weaves, and retaining their structural features.

The derivatives of <u>elementary weaves</u> include:
(1) Plain weave derivatives: <u>rib weaves</u>, <u>hopsack weaves</u>, etc;
(2) Twill weave derivatives: <u>reinforced twill</u>, <u>compound twill</u>, <u>angled twill</u>, <u>curved twill</u>, <u>waved or pointed twill</u>, <u>broken twill</u>, <u>diamond twill</u>, <u>zigzag twill</u>, <u>entwined twill</u>, etc;
(3) Satin/sateen derivatives: <u>reinforced satin/sateen</u>, <u>rearranged satin/sateen</u>.

原组织

重平组织/方平加强斜纹/复合斜纹/急斜纹/曲线斜纹/山形斜纹/破斜纹/菱形/锯齿/芦席斜纹/加强缎纹/变则缎纹/平纹变化组织

3.1 Plain weave derivatives

This group of structures comprises various simple weaves which are varieties of the plain weave and can be produced on two <u>heald shafts</u>. The extension of the plain weave can proceed either vertically, grouping together several picks in the same shed, which results in warp rib, or horizontally, with groups of neighboring ends working in tandem and producing weft rib, or in both directions simultaneously, resulting in mat, or <u>hopsack weave</u>.

Plain weave and its derivatives are often used for producing <u>fabric selvedges</u>.

片综

方平组织

布边

3.1.1 Rib weaves
Rib weaves are obtained by extending the plain weave in either warp or weft direction.

重平组织

3.1.1.1 Warp rib weaves
Warp ribs are constructed by inserting several picks in succession into the same shed of an ordinary plain weave. This forms a rib effect across the fabric. They are woven with a substantially higher number of ends than picks. The warp should cover the weft on both sides of the fabric. The weft yarn has usually less twist and is of <u>thicker count</u>(see Fig. 3.1).

经重平组织

粗支

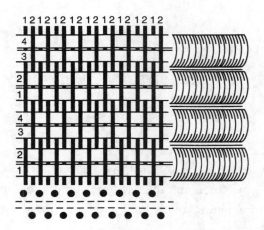

Fig. 3.1　Warp rib weave

正规经重平组织

• **Regular warp rib**

The same number of picks are inserted in each successive rib, giving the fabric a regular appearance.

It can consist of several picks in one shed depending on the count of the weft yarn and the width of the rib required.

The correct sett plays a very important part when constructing rib fabrics. Insufficient ends or picks render the fabric liable to slippage.

In modern looms warp rib can often be woven by inserting two picks simultaneously into the same shed, thus making a saving in the production costs. This applies also to hopsack and other similar weaves where two picks are inserted into the same shed.

经浮线/纬浮线

The warp rib is denoted by a fraction form. The numerator shows the number of warp floats and the denominator, the number of weft floats on the same thread within the repeat. Fig.3.2 shows a $\frac{2}{2}$ warp rib. Fig.3.3 shows a $\frac{3}{3}$ warp rib.

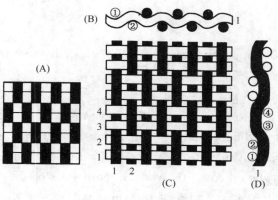

Fig. 3.2　Warp rib $\frac{2}{2}$

• Irregular warp rib

变化经重平组织

A variation in the width of rib is obtained by inserting different numbers of picks into each successive shed. See Fig. 3.4.

The warp rib weave diagram is drawn in following steps.

Fig. 3.3　Warp rib $\frac{3}{3}$　　　Fig. 3.4　$\frac{2}{1}$ irregular warp rib

(1) Calculate the weft repeat R_y: R_y = numerator + denominator,

$$R_0 = 2.$$

(2) Draw the first end according to the fraction given.

(3) Draw the second end opposite to the first one.

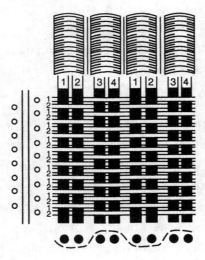

Fig. 3.5　Weft rib weave

3.1.1.2　Weft rib weaves

纬重平组织

Weft ribs are constructed with several warp threads used as one when interlacing with each pick in succession. They form a vertical rib in the fabric. They have a considerably higher number of picks than ends. The weft should cover the warp on either side of the fabric. Finer weft yarns will give better coverage and make it easier to achieve the required pick density. Weft rib, due to its high number of picks, increases the production costs.

• Regular weft rib

规则纬重平组织

An equal number of ends are used to form each rib, giving the fabric a regular appearance. The number of ends depends on the width of the rib required.

The weft rib is also denoted by a fraction form. In weft rib the sum of the fraction, numerator and denominator is equal to the warp repeat. Fig. 3.6 shows the $\frac{2}{2}$ weft rib at A and $\frac{3}{3}$ weft rib at B.

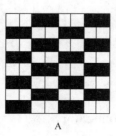

Fig. 3.6　Regular rib

Fig. 3.7　$\frac{2}{1}$ Irregular rib

- **Irregular weft rib**

A variation in the width of the rib is obtained by varying the number of ends in each successive rib, as shown in Fig. 3.7.

The weft rib weave diagram is drawn as following.

(1) Calculate the warp repeat R_0: R_0 = numerator + denominator,

$$R_y = 2.$$

(2) Draw the first pick according to the fraction given.

(3) Draw the second pick opposite to the first one.

It need to be noted that warp rib weaves produce ribs running weft-way, and weft rib weaves produce ribs running warp-way.

Rib gives a more flexible cloth than plain weave and has many applications. Fabrics are woven in silk, cotton, wool and man-made fibers. Their end uses range from dress fabrics, coats, suits, millinery, ribbons and wedding to upholstery and drapery.

3.1.2　Hopsack weaves

Hopsack weaves are constructed by extending the plain weave both vertically and horizontally. There are two or more warp threads working in the same manner and two or more weft threads grouped in the same shed. The hopsack weave is denoted by a fraction, the numerator is the numbers of warp floats, the denominator is that of weft floats on each thread. The sum of the numerator and denominator shows the repeat on warp and weft (Fig. 3.8).

- **Regular hopsack**

Most regular hopsack are woven with the same number of ends and picks

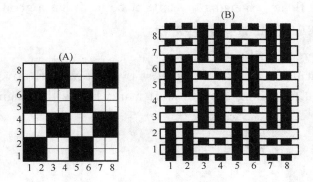

Fig. 3.8　Hopsack weaves

and the same yarn count. Equal warp floats exchange with equal weft floats. See Fig. 3.9.

- **Irregular hopsack**　　　　　　　　　　　　　　变化方平组织

Different units of hopsack are arranged in one repeat, with the distribution of warp or weft floats being equal or a predominance of either. See Fig. 3.10.

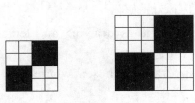 

Fig. 3.9　$\frac{2}{2}$, $\frac{3}{3}$ Regular hopsack　　Fig. 3.10　$\frac{3\ 2}{2\ 2}$ Irregular hopsack

The irregular hopsack diagram is drawn in following steps.

(1) Calculate the repeat：

　　$R_0 = R_y$ = sum of the numerator + sum of the denominator.

(2) Draw the first end and first pick depended on the fraction.

(3) Based on the first pick, draw the ends which have the same warp float as first end.

(4) Draw the other ends opposite to the first end.

Hopsack weave fabrics are less stiff than plain due to its fewer intersections, and they have smooth and lustrous surface. Hopsack fabrics are suitable for apparel, drapery, and are often used for fabric selvedge.

3.2　Twill weave derivatives　　　　　　　　　　　　斜纹变化组织

Variations of twill weaves are many. Twill weaves can be modified by extending the floats, changing the shift or both. It has a great potential for introducing <u>ornamentation</u> into fabrics.

3.2.1 <u>Reinforced twills</u>

Reinforced twill weaves are the simplest twill weave derivatives, which can be constructed by adding warp or weft marks beside the original ones. See Fig. 3.11.

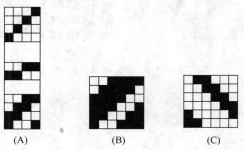

(A)　　　　(B)　　　　(C)

Fig. 3.11　Reinforced twills

The direction of the diagonal line is denoted by <u>arrowhead</u>. e.g. $\frac{2}{2}\nearrow$ at Fig. 3.11.(A), $\frac{4}{2}\nearrow$ at Fig. 3.11.(B), $\frac{2}{4}\nwarrow$ at Fig. 3.11(A)(C).

Drawing reinforced twill diagram is the same as drawing an elementary twill.

Straight draft is considered first. Skip draft is chosen where the warp density is too high. See Fig. 3.12.

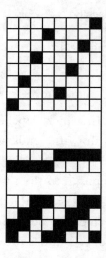

Fig. 3.12　Skip draft for reinforced weave

Reinforced twill weaves are widely used in varies fabrics, such as <u>serge</u>, <u>gabardine</u>, <u>drill</u>, and also used in selvedges of other fabrics.

3.2.2 Compound twills

The compound twill is obtained by constructing two or more parallel twill lines in the same area. It has a fancy diagonal appearance.

The compound twill is denoted by a fraction form. The numerator indicates warp floats, and the denominator indicates weft overlaps. The direction of the diagonal line is denoted by arrowhead. See Figs. 3.13 (A), (B), (C).

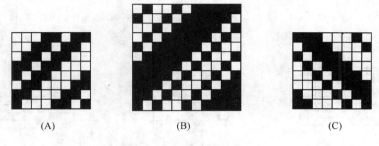

(A)　　　　　　　(B)　　　　　　　(C)

Fig. 3.13　Compound weave

Fig. 3.13 (A) shows compound twill $\frac{2\ 2}{1\ 3}\nearrow$; Fig. 3.13 (B) shows compound twill $\frac{5\ 1\ 1}{1\ 2\ 1}\nearrow$, and Fig. 3.13 (C) — compound twill $\frac{3\ 1}{3\ 1}\nwarrow$.

Compound twills are drawn in the following steps:

(1) Calculate the repeat R, $R_0 = R_y$ = numerator + denominator.

(2) Draw the first end according to the fraction given.

(3) Draw the other ends based on the first one and shift. (Here shift equals to ±1).

Straight draft is more often used for compound twills.

Compound twills are widely used in fancy fabrics due to its interesting appearance.

3.2.3 Elongated twills

In practice, the relationships between the warp density P_0 and weft density P_y: $P_0 = P_y$; $P_0 > P_y$ and $P_0 < P_y$.

The density ratio changes the fabric appearance.

A weave drawn on paper will have the form of a square when $P_0 = P_y$. In case of other ratios the weave diagram will assume the form of a rectangular.

When $P_0 = P_y$ the twill diagonal on the fabric will make an angle of 45° with the horizontal line. The tangent of the inclination angle of twill diagonal to the horizontal plane is

$$\tan \alpha = \frac{1}{P_y} \Big/ \frac{1}{P_0} = P_0/P_y.$$

The inclination angle depends on the density ratio:

$$P_0 = P_y, \tan \alpha = 1, \alpha = 45°;$$
$$P_0 < P_y, \tan \alpha < 1, \alpha < 45°;$$
$$P_0 > P_y, \tan \alpha > 1, \alpha > 45°.$$

It can be seen that the inclination angle of the diagonal changes when the ratio of density varies. This fact may be used for changing the appearance of the fabric. See Fig. 3.14.

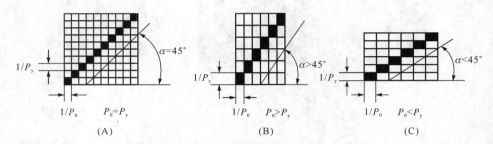

Fig. 3.14 Elongated weaves

To increase the inclination angle of the twill, we can increase the warp density, but the ratio P_0 to P_y should not be higher than 2. Because of considerations of physical and mechanical properties of the fabric. If we want to increase the inclination angle of the twill more, we can also change the type of weave. The inclination angle of the twill increases if we choose a larger vertical shift. Practically, the shift can be 2 or 3.

The increase of the shift from 1 to 2 is equivalent to the doubling of warp-density.

Twill constructed with an increased shift is called elongated twill. There are two elongated twills, <u>diagonal weave</u> and <u>reclining twill</u>.

急斜纹/缓斜纹

Twill, constructed with an increased vertical shift, is called diagonal weave.

For the diagonal weave

$$\tan \alpha = P_0 S_0 / P_y,$$

while constructing the diagonal twill two cases may occur.

(1) The repeat of the basic twill can be divided by the increased shift.

(2) The repeat of the basic twill cannot be divided by the increased shift. The warp repeat of the diagonal weave

$$R_0 = R/S_0.$$

If the repeat of the basic twill can not be divided by the increased shift (S_0) the diagonal weave will have the repeat $R_0 = R_y = R$.

The weft repeat

$$R_y = R.$$

Example 1. Construct a diagonal weave on the basis of twill $\frac{4\ 4\ 1}{1\ 2\ 2}$, with $S_0 = 2$ (Fig. 3.15).

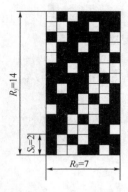

Fig. 3.15 Diagonal weave, $R_0 = 7$; $R_y = 14$

According to the above formulae, $R_y = (4+1+4+2+1+2) = 14$; $R_0 = \frac{R_0}{R_y} = \frac{14}{2} = 7$.

Example 2. Construct a diagonal weave on the basis of twill $\frac{5\ 1\ 1}{1\ 2\ 1}$, with $S_0 = 2$. In this case $R_y = (5+1+1+2+1+1) = 11$, since $R_0 = \frac{11}{2}$ cannot be integrally divided, $R_0 = R_y = 11$. The diagonal weave constructed on the basis of twill $\frac{5\ 1\ 1}{1\ 2\ 1}$ is shown in Fig. 3.16.

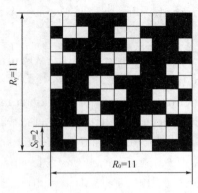

Fig. 3.16 Diagonal weave $R_0 = 11$, $R_y = 11$

The procedure of drawing diagonal weave diagrams can be summarized as follows:

(1) Calculate the repeats, R_0 & R_y.

$$R_0 = \frac{\text{basic twill repeat}}{\text{the highest common factor of basic twill repeat and shift}}$$

i.e. $R_0 = \dfrac{R_{OB}}{\text{HCF of } R_{OB} \,\&\, S_0}$

$$R_Y = R_{YB}$$

Here, R_{OB}, R_{yB} are basic twill repeats.

(2) Draw the first end according to the formula of the basic twill.

(3) Draw the other ends according to the same formula, but with the changed shift.

Twill, constructed with an increased horizontal shift, is called reclining twill.

For the reclining weave

$$\tan \alpha = \frac{P_0}{P_y \cdot S_y}.$$

The weave diagram drawing of reclining twills is similar to drawing diagonal weaves. Figure 3.17 shows the $\dfrac{4\ 1\ 1}{2\ 2\ 2}$ with reclining weave $S_y = 2$.

Fig. 3.17 Reclining twill weave

Some fabrics made from the diagonal weave are: whipcord, lady's cloth.

3.2.4 Curved twills

Curved twills produce a curvilinear appearance when the inclination angle of diagonal changes gradually.

As we know, the inclination angle of the diagonal may be changed by changing the shift. The curved twill can be constructed by using varying shifts of the basic twill. An increase of the shift makes the inclination angle of the twill diagonal bigger and a decrease of the shift makes the inclination angle smaller. The curved twill has a weft repeat equal to the weft repeat of the basic twill, but the warp repeat may be bigger, depending on the specification of the series of shifts.

To construct a curved twill, we must pay attention to the following two points.

(1) The sum of shifts must equal 0 or can be divided by the weft repeat of the basic twill without a remainder.

(2) The biggest shift must be less than the biggest float length in order to avoid twill line breakage.

The curved twill may be constructed using varying shifts on the weft. In this case, the warp repeat of the curved twill equals the repeat of the basic twill.

Fig. 3.18 shows a curved twill based on two repeats of the $\frac{4\ 1\ 1}{3\ 1\ 3}$ twill.

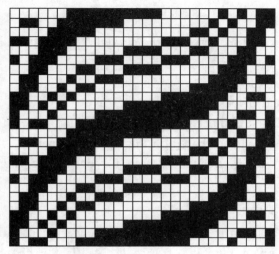

Fig. 3.18　Curved weave

$S_0 = 2, 2, 2, 1, 1, 1, 0, 1, 0, 0, 1, 0, 0, 0, 1, 0, 0, 1, 0, 1, 1, 1, 1, 2, 2, 2, 2$ is constructed. The sum of shifts equals 26. The basic twill is twill $\frac{4\ 1\ 1}{3\ 1\ 3}$. The weft repeat of this twill is 13. The sum of shifts 26 is divided by a repeat, 13 without a remainder. Because of the number of shifts used for this weave, the repeat of curved twill is 28. It should be pointed out that the lifting plan is similar to that of the basic twill. The curved draft is used in this case.

More <u>aesthetic appearance</u> can be achieved in Fig. 3.19 if the curved twill image is mirrored.　　　　　　　　　　　　　　　　　　　　　漂亮的外观

The basic twill is still twill $\frac{4\ 1\ 1}{3\ 1\ 3}$. The shifts are $2, 2, 1, 1, 0, 1, 0, 0, 1, 1, 0, 1, 1, 1, -1, -1, -1, 0, -1, -1, 0, 0, -1, 0, -1, -1, -2, -2$.

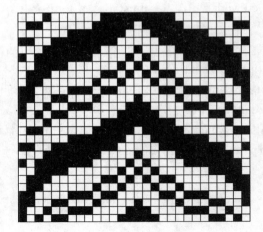

Fig. 3.19 Curved weave

The procedure used for drawing the curved twills is given below.

(1) Calculate the repeat:

R_0 = number of the shifts,

R_y = weft repeat of basic twill.

(2) Draw the first end according to the formula of the basic weave.

(3) Draw the other ends based on the shifts designed.

Curved twill weaves are widely used for <u>decorative fabrics</u> and apparel fabrics.

3.2.5 Angled twills

The angled twill is constructed by changing the sign of shift from plus to minus, after a given number of threads. With the change of sign, the direction of the twill line is changed.

Angled twill weave can be divided into <u>vertical angled twill</u> and <u>horizontal angled twill</u>.

- **Vertical angled twill**

The vertical angled twill has two diagonal lines which are symmetrical about a predetermined warp thread.

To construct the angled twill, we calculate its repeat first. The weft repeat is equal to the repeat of the basic twill, i.e. $R_y = R_{yb}$. The warp repeat can be determined by the following equation

$$R_0 = 2K_0 - 2,$$

where K_0 is the number of warp threads after which the sign of shift changes.

Example 1. Construct the angled twill on the basis of twill $\frac{2}{4}$ (see

Fig. 3. 20). The repeat of basic twill is 6. The repeat of angled twill is calculated as follows:

$$R_y = R_{yb} = 6,$$
$$R_0 = 2K_0 - 2 = 2 \times 6 - 2 = 10.$$

For producing this weave, the point draft is used.

Example 2. Construct the angled twill on the basis of twill $\frac{3\ 1}{2\ 2}$, $K_0 = 8$. See Fig. 3. 21.

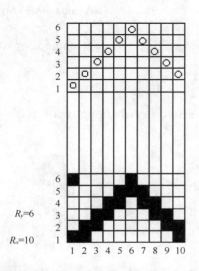

Fig. 3. 20 Angled twill and point draft   Fig. 3. 21 Vertical angled twill

(1) Calculate the repeat:

$$R_0 = 2K_0 - 2 = 2 \times 8 - 2 = 14,$$
$$R_y = R_{yb} = 8.$$

(2) Draw the ends from 1 to K_0 based on the twill $\frac{3\ 1}{2\ 2}$.

(3) Draw the others based on the changing sign of shift.

• **Horizontal angled twills**

Example 3. Construct the angled twill on the basis of twill $\frac{1\ 2\ 2}{1\ 1\ 1}$ by changing the sign of horizontal shift after K_y weft threads. K_y is the number of weft threads after which the sign of shift changes. Here suppose $K_y = 8$.

The repeat of weave is calculated as follows:

$$R_0 = 8.$$

$R_y = 2K_y - 2 = 2 \times 8 - 2 = 14$. The <u>angled twill</u> is constructed in Fig. 3. 22.　角度斜纹
Note that short warp or weft floats should be used so as to avoid overlong

floats when the weave is reversed. Vertical angled twill weaves are achieved with point draft. Horizontal angled twills are achieved with straight draft. Vertical angled twills are widely used for fabrics such as lady's dressing, woolen coat, sheet, and various tweed.

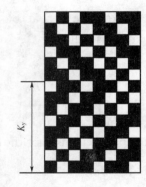

Fig. 3.22 Horizontal angled twill

3.2.6 Herringbone and broken twills

The twill line of herringbone weaves is broken at predetermined intervals to continue in the opposite direction.

Cuts or breaks occur in the fabric where warp floats are exchanged for weft floats and vice versa.

Clear cuts avoid long floats that can occur in angled weaves.

The base weave, $\frac{2}{2}$ twill is used to form two herringbone twills as shown in Fig. 3.23.

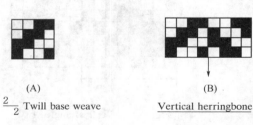 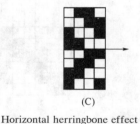

(A) $\frac{2}{2}$ Twill base weave　　(B) Vertical herringbone　　(C) Horizontal herringbone effect

Fig. 3.23 Herringbone weave

The base weave $\frac{3\ 1\ 2}{3\ 2\ 1}$ with $K_0 = 6$ is used to form the herringbone twill shows in Fig. 3.24.

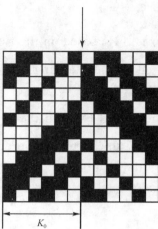

Fig. 3.24 Vertical herringbone weave

The base weave $\frac{1\ 3\ 1}{1\ 1\ 3}$ twill with $K = 10$ is used to form the herringbone twill shown in Fig. 3.25.

Herringbone weave construction is similar to that for angled twill except for the repeat calculation.

For vertical herringbone: $R_0 = 2K$, $R_y = R_{yb}$.

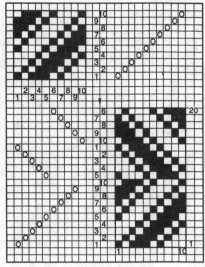

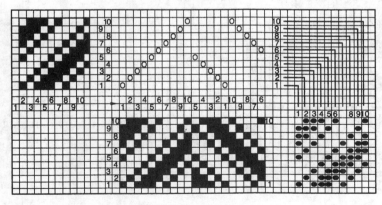

Fig. 3.25

For horizontal herringbone: $R_0 = R_{0b}$, $R_y = 2K$

A broken twill is formed by a break in the continuation of the twill line at predetermined intervals(see Fig. 3.26).

The 4-end broken twill is often used for various fabrics (See Fig. 3.27).

Fig. 3.26 Broken twill

With a $\frac{2}{2}$ base twill alternate ends or picks interlace in plain weave. A clear broken twills is achieved.

Further variations can be made by extending the repeat to create larger effects.

In these examples the base twill has been rearranged and extended to three times the number of picks or ends of the original repeat.

In the weft direction the sequence of the picking order is extended from 6 to 18 picks.

In the warp direction a stepped draft replaces the straight draft. The repeat is extended from 6 to 18 ends.

● (lifts) indicated the sequence of the rearranged ends or picks. They correspond also with the draft and weft interlacing.

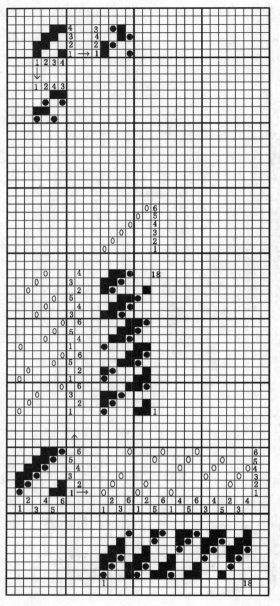

Fig. 3.27 Various broken twills

Broken drafts are suitable for herringbone and broken twills. Compared with fabrics using angled twills. Fabrics from herringbone twill given a better appearance to the fabric and are a more stable structure. They are used for herringbone tweeds, sheets, and various decorative fabrics.

3.2.7 Diamond and diaper

Diamond designs are based on angled twill and can be considered as a combination of a vertical angled twill and a horizontal angled twill. They are constructed in the following steps (see Fig. 3.28).

(1) Choosing a base weave: here $\frac{2}{2}$ twill is chosen as base weave.

(2) Determining K_0, K_y: here, in Fig. 3.28, $K_0 = 4$, $K_y = 4$.

(K_0 and K_y may or may not be equal to each other). So the repeat $R_0 = 2K_0 - 2 = 6$; $R_y = 2K_y - 2 = 6$. We can draw the outline of the weave.

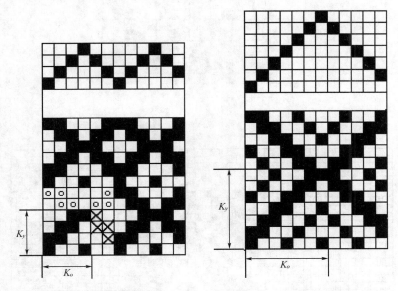

Fig. 3.28　Diamond weave　　Fig. 3.29　Diamond weave

(3) Drawing the basic part according to the base weave and K_0, K_y. In Fig. 3.28 this is marked as "■".

(4) Drawing the angled twill. In Fig. 3.28, this is marked as "⊠".

(5) Drawing the other parts of the weave by taking the weft thread K_y as the axes of symmetry. In Fig. 3.28, this is marked as "⊡".

Fig. 3.29 shows a diamond weave based on $\frac{2\ 1}{2\ 2}$ ↗ $K_0 = K_y = 7$.

Diaper designs are based on herringbone twill and can be considered as a vertical herringbone twill combining with a horizontal herringbone twill. See Fig. 3.30(A), (B).

The repeat of a diaper weave is divided into four equal quarters. The base twill is entered into the first quarter; for the next and every following quarter warp floats are exchanged for weft floats and vice verse.

Diamond and diaper weaves are suitable for pointed draft.

Diamond and diaper weaves are widely used for lady's cloth, sheet fabrics and some woollen fabrics.

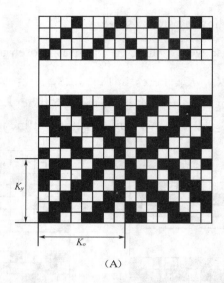

(A)

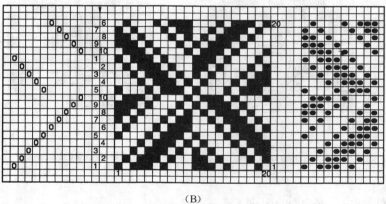

(B)

Fig. 3.30   Diaper weaves

锯齿斜纹

3.2.8 Zigzag weaves

This is a variation of the angled twill.

In an angled twill all waves are on the same level. Whereas in a zigzag weave, the position of the wave are arranged in an ascending or descending diagonal line. Each top point moves a number of threads which is called the zigzag move. They are constructed in following steps (see Fig.3.31).

(1) Choose base weave. Here, $\frac{2\ \ 1}{1\ \ 2}$ twill is chosen as a base weave.

(2) Determine K_0 which is the number of threads before changing direction. Here, $K_0 = 9$. Choose the zigzag move = 4. We can calculate the repeats based on the parameter above.

$$\text{The ends of each zigzag} = (2K_0 - 2) - m \ (m \text{ — the zigzag move})$$
$$= 2 \times 9 - 2 - 4 = 12.$$

UNIT Ⅰ General Knowledge and Simple Construction   47

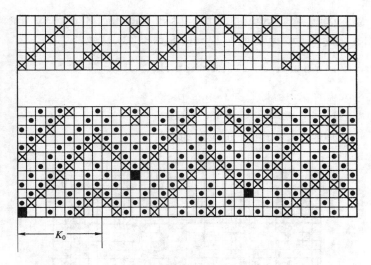

Fig. 3.31 Zigzag weave

The number of zigzage = $\dfrac{\text{base weave repeat}}{\text{HCF of base weave repeat and zigzag move}}$ 最大公约数

$= \dfrac{6}{2} = 3.$

So repeat R_0 = the number of zigzag × the ends of each zigzag
$= 3 \times 12 = 36$,

R_y = base weave repeat = 6.

(3) Draw the outline of each zigzag, and then draw the first interlacing point at the first end. See Fig.3.30 mark "■".

(4) Draw the warp threads from 1 to K_0 according to base weave, and draw the other threads in the opposite direction until the last end of the first zigzag.

(5) Draw the other zigzag in the same way.

The zigzag weaves are used for apparels, sheets and decorative fabrics.

3.2.9 Entwined twills   芦席斜纹

This is a variation of the broken twill. It is generally developed from a combination of even sided Z and S twills and gives the fabric a simulated lattice appearance.   窗格外观

The repeat of an entwined twill is always a multiple of the base twill, e.g.: 2/2 twill has a repeat of 4 ends/4 picks, and the new repeat can be: $4 \times 2 = 8$, $4 \times 3 = 12$, $4 \times 4 = 16$ etc.

Each repeat consists of an equal number of Z and S lines. The repeat is therefore divided into two halves.

Details for an entwined twill are calculated as follows:

(1) New repeat ÷ base twill repeat = number of twill lines in one repeat.

(2) New repeat ÷ 2 = number of ends in one line (length of line);
Examples are shown in Fig. 3.32 and Fig. 3.33.

$\frac{2}{2}$ base twill repeat: 2 + 2 = 4 ends/picks,

new repeat: 4 × 2 = 8 ends/picks.

(a) 8 ÷ 4 = 2 lines in one repeat,

(b) 8 ÷ 2 = 4 ends, length of line.

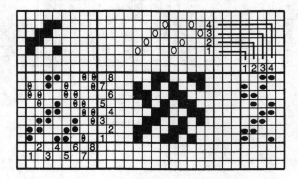

Fig. 3.32 Entwined twill

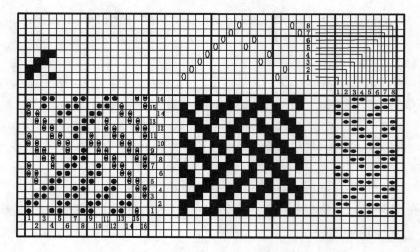

Fig. 3.33 Entwined twill

$\frac{2}{2}$ base twill repeat: 2 + 2 = 4 ends/picks,

new repeat: 4 × 4 = 16 end/picks.

(a) 16 ÷ 4 = 4 lines in one repeat,

(b) 16 ÷ 2 = 8 ends, length of line.

See Fig. 3.34. $\frac{3}{3}$ base twill repeat: 3 + 3 = 6 ends/picks,

new repeat: 6 × 3 = 18 ends/picks.

(a) 18 ÷ 6 = 3 lines in one repeat,

(b) 18÷2=9 ends, length of line.

Entwined twill weaves can be constructed in the following steps.

(1) Design the size of pattern, and calculate the repeats, lines in one repeat, length of the line as mentioned above.

(2) Divide the repeat into two parts equally. Draw the first Z twill line according to the base weave in the first part. See Fig.3.34 mark "■".

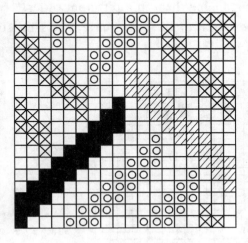

Fig. 3.34 Entwined twill

(3) Draw the first S twill lines by stepping up the base weave float. In Fig.3.34 this is marked as "▨".

(4) Draw the other Z twill lines in same length in the first part. In Fig.3.34 this is marked as "◯".

(5) In the same way, draw the other S twill lines in the second part. In Fig.3.35 this is marked as "⊠".

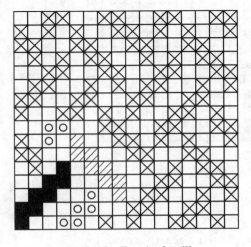

Fig. 3.35 Entwined twill

Figure 3.35 shows another entwined twill, which is based on $\frac{2}{2}$ with 2 twill lines in each direction, and 4 repeats of this entwined weave are shown. Entwined twills are used for apparels and sheets.

变则斜纹
重排列

3.2.10 Rearranged twills

<u>Rearranging</u> a weave means taking single threads or groups of threads of the base weave and rearranging them in a different order. If the rearrangement does not exceed the repeat of the base weave, the same straight draft can be used(See Fig. 3.36).

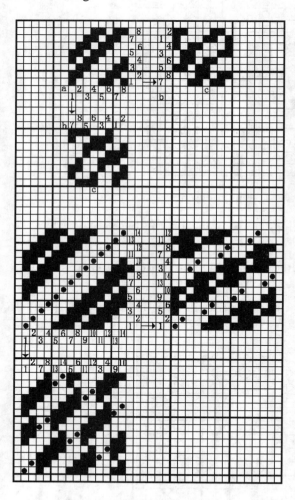

Base twill $\frac{2\ \ 2}{1\ \ 3}$

The picks are rearranged in groups of two.

Numbers identify the groups of picks that have been exchanged.

The ends are rearranged in groups of two.

Both number systems are identical.

Base twill $\frac{1\ \ 3\ \ 3}{3\ \ 1\ \ 3}$

The picks are rearranged in groups of two.

The points of the base line have been arranged in groups of two in sateen.

A different rearrangement has been applied in warp direction.

Fig. 3.36 Rearranged twills

This type of weave can be constructed as follows.
(a) Select a base weave.
(b) Determine a plan for rearrangement.
(c) Rearrange the threads: ends or picks.

3.2.11 Shaded twill

The shaded twill represents a gradual transition from the twill with weft effect to the twill with warp effect, and vice versa. It is used mostly in Jacquard weaving for large-pattern fabrics.

The repeat of warp and weft for the shaded twill is determined by the formula

$$R_0 = R_{0b}(R_{0b} - 1),$$

where: R_0— new weave repeat;

R_{0b}— the base weave repeat.

In case of shaded twill the straight draft is used. Fig. 3.37 is a shaded twill transiting from the $\frac{1}{4}$ twill to the $\frac{4}{1}$ twill.

Fig. 3.37 Shaded twill

3.3 Derivatives from satin/sateen weaves

In simple derivatives, a new design is built up by using the original sateen as the base and adding floats, as required to each base interlacement.

Reinforced sateen. The way of obtaining the reinforced sateen is similar to that of reinforced twill.

If it is necessary to construct the reinforced sateen on the basis of simple sateen $\frac{8}{5}$. One more warp floats should be added to every float of the basic sateen (Fig. 3.38). This kind of weave is widely used in cotton weaving. Additional floats in the repeat make the fabric structure stronger.

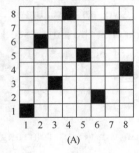 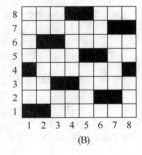

(A) (B)

Fig. 3.38 Reinforced sateen

Figure 3.39 shows a $\frac{11}{8}$ reinforced sateen which is often used for worsted fabric called <u>satin back gabardine</u>. The face of the fabric has a <u>twill appearance</u> and back has a <u>satin appearance</u>.

缎背华达呢/
斜纹外观/缎
纹外观

<u>Shaded sateen</u>. Like the shaded twill the shaded satin represents a gradual transition from the sateen with weft effect to the satin with warp effect, and vice versa.

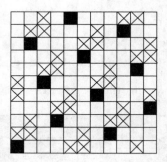

Fig. 3.39 Reinforced sateen

The repeat R_y and R_0 of the shaded sateen is determined by the formula:

$$R_y = R_{yb} = 5,$$
$$R_0 = R_{0b}(R_{0b} - 1) = 5(5 - 1) = 20,$$

where: R_{0b}, R_{yb}—— base weave repeat;
R_y, R_0—— new weave repeat.

The shaded sateen $\frac{5}{3}$ is illustrated in Fig. 3.40.

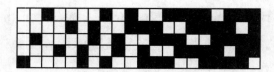

Fig. 3.40 Shaded sateen

不规则外观

Rearranged sateen/satin weaves can give an <u>irregular appearance</u> by rearranging the sequence of the picks or varying the shift. See Fig. 3.41, and Fig. 3.42. Fig. 3.41 indicates the exchange of picks.

Fig. 3.42 shows a 6-end <u>rearranged sateen</u>.

Shifts $(S_y) = 4, 3, 2, 2, 3.$

The same effect can be achieved for satin weaves by rearranging the ends.

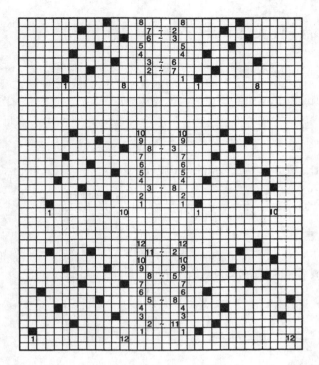

Fig. 3.41 Rearranged sateen

Fig. 3.42 Rearranged sateen

HOMEWORKS

1. Drawing the weave diagrams

    (1) $\frac{2}{2}$ warp rib;

    (2) $\frac{2}{2}$ weft rib;

    (3) $\frac{2}{2}$ hopsack;

    (4) $\frac{3}{3}$ hopsack;

    (5) irregular warp rib $\frac{2}{1}\frac{2}{2}$;

    (6) $\frac{3}{1}$ irregular weft rib,

(7) irregular hopsack $\frac{3\ 2}{3\ 2}$;

(8) irregular hopsack $\frac{1\ 2}{2\ 1}$.

2. Drawing the following reinforced twill weaves.

$$\frac{2}{3}\nearrow,\ \frac{4}{3}\nwarrow,\ \frac{3}{5}\nearrow,\ \frac{4}{4}\nwarrow$$

3. Drawing the following compound twill weaves.

$$\frac{2\ 3}{3\ 2}\nearrow,\ \frac{1\ 2}{2\ 3}\nwarrow,\ \frac{2\ 2}{1\ 2}\nwarrow,\ \frac{3\ 1\ 1}{2\ 2\ 1}\nearrow,\ \frac{4\ 2\ 3}{2\ 1\ 3}\nearrow,$$

$$\frac{1\ 2\ 3\ 4\ 3\ 2}{4\ 3\ 2\ 1\ 2\ 1}\nearrow$$

4. Whipcords made from weave formula:

(1) $\frac{7\ 1\ 1\ 1\ 1\ 1}{1\ 2\ 1\ 1\ 2\ 1}$, $S_0 = 2$

(2) $\frac{5\ 1\ 1}{1\ 2\ 1}$, $S_0 = 2$

drawing the above weaves.

5. Drawing the curved weave, the basic twill is $\frac{4\ 1\ 1}{3\ 1\ 3}$, the vertical shift $S_0 = 2, 2, 1, 1, 0, 1, 0, 0, 1, 1, 0, 1, 1, 1, -1, -1, -1, 0, -1, -1, 0, 0, -1, 0, -1, -1, -2, -2$, and drawing the weaving plan using as small number of shafts as possible.

6. Drawing the following angled twills, and adding the weaving plan:

(1) basic weave $\frac{2\ 1}{1\ 2}\nearrow$, $k_0 = 9$;

(2) basic weave $\frac{1\ 1\ 3}{1\ 2\ 2}\nearrow$, $k_0 = 10$;

(3) basic weave $\frac{2\ 1}{3\ 2}\nearrow$, $k_0 = 12$.

7. Drawing the following Herringbone weaves:

(1) basic weave $\frac{4}{4}\nearrow$, $k_0 = 8$;

(2) basic weave $\frac{4\ 1\ 1}{1\ 4\ 1}\nearrow$, $k_0 = 12$.

8. Drawing the diamond weave, the basic weave is $\frac{2}{2}\nearrow$, $K_0 = 10$, $K_y = 10$

9. Drawing the zigzag weave:

base weave $\frac{2\ 1}{3\ 2}\nearrow$, $K_0 = 8$, $m = 4$.

10. Drawing the following entwined weaves:

(1) Base twill $\frac{2}{2}\nearrow$, new repeat 16;

(2) Base twill $\frac{3}{3}\nearrow$, twill lines 3;

(3) Base twill $\frac{4}{4}\nearrow$, twill lines 2.

11. Base weave $\frac{1}{3}\frac{3}{1}\frac{3}{3}\nearrow$, the order of warp threads is 1, 2, 7, 8, 13, 14, 5, 6, 11, 12, 3, 4, 9, 10, draw the rearranged twills.

12. Drawing a shaded twill from $\frac{1}{3}\nearrow$ to $\frac{3}{1}\nearrow$.

Chapter Four
Combined Weaves
联合组织

Fabric woven with elementary weaves usually has a smooth surface. The plain weave has an even and regular surface. On the face of twill fabrics there are distinct diagonal lines. The face of sateen weave fabric is smooth and lustrous due to uniform arrangement of single warp floats.

Contrary to elementary weaves the combined weaves produce irregular or uneven fabric surface or small woven figures on the fabric. These weaves are constructed on the basis of two or more elementary weaves and their derivatives.

Combined weaves may be divided into the following groups: stripe and check weaves, crepe weaves, mock leno weaves, huckaback, honeycomb weaves, Bedford weaves, and distorted weaves.

4.1 Stripe and check weaves

Some fabrics are made with longitudinal and cross stripes of different widths and different weaves on the surface.

Combination of longitudinal and cross stripes on the fabric forms checks in places of their intersection.

These weaves are constructed on the basis of two or more elementary weaves or their derivatives.

The methods of calculation and construction of such fabrics are given below.

4.1.1 Calculation and construction of stripe weaves

Calculation and construction of fabric with longitudinal stripes. In this case, the warp repeat depends on the number of stripes, the width of each stripe, the warp density, and the stripe weave.

To calculate the number of warp threads, the warp density is multiplied by the width of each stripe. The numbers received are corrected to get a whole number of the repeats in each stripe.

To produce a fabric with longitudinal stripes, a grouped draft is used. It is convenient to have a whole number of repeats in each group of heald shafts in the case of grouped draft. The number of heald shafts equals the sum of heald shafts necessary for producing each stripe separately.

If, for example, a fabric contains three stripes, and the width of the 1st stripe is A cm, the 2nd B cm, and the 3rd C cm, then the warp repeat of the fabric is, approximately,

$$R_0 = P_0 A + P_0 B + P_0 C, \quad R_0 = P_0(A + B + C),$$

here P_0 is the warp density measured in ends/cm.

The weft repeat equals the <u>least common multiple</u> of the weft repeat of the stripe weaves.　　最小公倍数

Example. Calculate for the fabric with two longitudinal stripes if the warp density is 22 threads/cm, the width of the 1st stripe is 1.4cm, the width of the 2nd stripe is 1.0cm. The weave for the first stripe is the twill $\frac{1}{2}$, and that for the 2nd stripe is the 4-shaft irregular sateen.

First of all, we determine the number of threads of the 1st stripe, which is $n_{01} = P_0 \times \text{width} = 22 \times 1.4 = 30.8$ threads.

The number received must be divided by the warp repeat of the 1st stripe weave, i.e. we <u>take</u> 30 as the number of threads of the 1st stripe. In the　取 same way we determine the number of threads of the 2nd stripe:

$$n_{02} = 22 \times 1.0 = 22.$$

We take it as 24 threads. The warp repeat of the fabric then is

$$R_0 = 30 + 24 = 54.$$

The weft repeat of the fabric designed in the least common multiple of 3 and 4, i.e. $R_y = 12$.

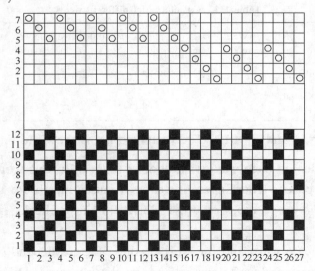

Fig. 4.1　**Weave with longitudinal stripe**

Thus, there are 54, warp threads in the repeat, but only 27 threads are shown in Fig. 4.1, i.e. 15 threads of the stripe A and 12 threads of the stripe B. Therefore, in the process of drawing-in the number of threads of these stripes should be repeated twice to get, respectively, 30 and 24 threads in the stripes.

In the draft, there should be 3 heald shafts to produce the twill $\frac{1}{2}$ of the 1st stripe, and 4 heald shafts to produce the sateen for the 2nd stripe. Thus, this fabric can be produced by using 7 heald shafts with a grouped draft.

Calculation and construction of fabric with cross stripes. The number of threads in the weft repeat depends on the number of stripes, width of each stripes, weft density, and stripe weaves.

To calculate the weft repeat, the weft density is multiplied by the width of each stripe. The number received is corrected to get a whole number of weft repeat in each stripe.

If, for example, the width of stripes are a, b, c, and the weft density is P_y threads/cm, then the weft repeat is

$$R_y = P_y(a + b + c).$$

The value of R_y need to be modified according to the weave repeats.

The warp repeat equals the least common multiple of the warp repeats of stripe weaves. For producing fabrics with cross stripes, straight draft is used. The number of heald shafts equals the warp repeat of the produced fabric.

Example. Using the weaves of the previous example, construct the fabric with 2 cross stripes. Weft density P_y = 23th./cm, and width a = 2cm, b = 1.5cm. Calculation is done in the form of Table 4.1.

Table 4.1 Calculation of different weaves

Stripe	Weave	Weft repeat	Width /cm	P_y /(th·cm^{-1})	Calculated number of threads	Accepted number of threads
1st(a)	Twill 1/2	3	2	23	23×2=46	45
2nd (b)	Irregular sat. with 4 shafts	4	1.5	23	23×1.5=34.5	36

Thus, R_y = 81 and R_0 = 12. The weaving plan is shown in Fig. 4.2.

4.1.2 Calculation of fabrics with checks from different weaves

Fabrics with checks from different weaves are used mostly for tablecloth and handkerchiefs. Some types of dress fabric also have checks from different stripes.

Repeats on the warp and the weft of these fabrics depend on the size of checks, the warp and weft density, and the check weaves. Different types of

UNIT Ⅰ   General Knowledge and Simple Construction   59

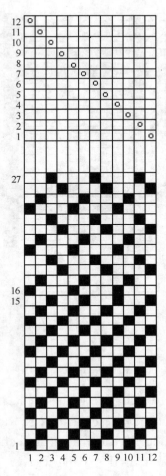

Fig. 4.2   Weave with cross stripes

weave can be combined in checks.

Example. Calculate for the handkerchief fabric with the check pattern given in Fig. 4.3.

The following must be determined: the warp repeat, the weft repeat, the number of heald shafts and the type of draft. The following characteristics are given:

cross stripe sizes: f, g, h, i, j;

longitudinal stripe size: a, b, c, d, e;

density on the warp and weft: P_0, P_y;

weaves of the background and the stripe.

Two features can be noticed while studying this pattern.

(1) The fabric is a structure with longitudinal stripes a, c, e and b, d. The longitudinal stripes (c, e) have the same structure as stripe a. It means that this fabric can be produced like any other fabrics with longitudinal

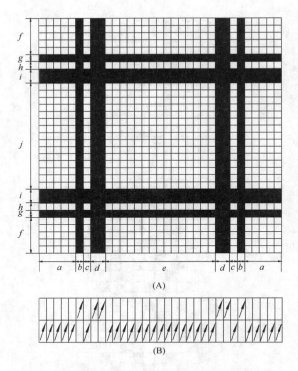

Fig. 4.3 Fabric with checks of different weaves

stripes, using grouped draft with two groups of heald shafts.

(2) The structure of stripes a, c, e is a fabric with cross stripes. The number of shafts for producing such a fabric equals to the least common multiple of the warp repeats of the stripe weave. The shafts for producing this stripe form one group.

The whole number of shafts equals the sum of shafts for the stripe and background.

The warp and weft repeats are determined by multiplying the respective densities by the stripe width; the number received must be corrected to get a whole number of elementary repeats in each stripe.

Let us calculate for the handkerchief according to the given widths of stripes of the model in Fig. 4.3.

$$a = 7.5\text{cm}, \ d = 3.0\text{cm}, \ g = 1.5\text{cm}, \ j = 24\text{cm},$$
$$b = 1.5\text{cm}, \ e = 24\text{cm}, \ h = 1.5\text{cm}, \ P_0 = 21.4\text{th/cm},$$
$$c = 1.5\text{cm}, \ f = 7.5\text{cm}, \ i = 3.0\text{cm}, \ P_y = 19.5\text{th/cm}.$$

The weave of the background is plain and of the border, a 4-shaft sateen.

Calculation of warp repeat. Now we examine the stripe (see Fig. 4.3). The stripe has the structure of a fabric with cross stripes. The warp repeat of this stripe equals the least common multiple of the warp repeats of the plain and sateen weaves ($R = 4$). The number of threads in the stripe equals the warp density multiplied by the stripe width, i.e.

$$P_0 a = 21.4 \times 7.5 = 160.5,$$

Let us correct this number to 160; 160/4 = 40 elementary repeats can be placed within stripe a. Let us consider the stripe b. The elementary warp repeat in this stripe is 4. The number of threads $P_0 b = 21.4 \times 1.5 = 32.1$. This number is corrected up to 32. The elementary repeat can be placed within the stripe 8 times.

The number of threads in the stripe d is twice as much as in the stripe b, i.e. $32 \times 2 = 64$. Because of the symmetry of the pattern, the warp repeat of the fabric designed is $(160 + 32 + 32 + 64) \times 2 + 480 = 1\,056$.

The type of drafts is the divided one using 8 heald shafts: 4 heald shafts for border, and 4 for background.

The draft is shown in Fig. 4.3(B) in an abridged form. The following number of groups of warp threads are shown in stripes: $a - 5$, $b - 1$, $c - 1$, $d - 2$, $e - 15$, and the total number is $5 + 1 + 1 + 2 + 15 + 2 + 1 + 1 + 5 = 33$. Four threads are presented in each group and totally $4 \times 33 = 132$ threads are given. Thus, in the process of drawing each group should be repeated 8 times to draw $132 \times 8 = 1\,056$ threads.

Calculation of weft repeat. Let us consider the stripe f. This stripe has the structure of a fabric with cross stripes. The elementary weft repeat of this stripe, equals the least common multiple of weft repeat of the stripe weaves. The weft repeats equal 2 and 4. The elementary weft repeat of the stripe f equals 4. The number of threads in the stripe equals $R = P_y \times f = 19.5 \times 7.5 = 146.25$.

Thus, 148/4 = 37 repeats can be placed within this stripe.

Now we consider the stripe g. The number of threads $R_y = P_y \times g = 19.5 \times 1.5 = 29.25$, but the weft repeat of the elementary weave is 4, that is why this number is corrected up to 28.

Let us consider the stripe i. The number of threads in the stripe is greater than in the stripe g; $R = 19.5 \times 3 = 58.5$ threads, but it is assumed 60.

As after the middle the pattern repeats itself, the weft repeat of the fabric designed equals $148 + 28 + 28 + 60 + 468 + 60 + 28 + 28 + 148 = 996$ threads.

4.2 Crepe weaves

绉组织

Crepe fabrics are characterized by a <u>pebbly</u> or <u>crinkled</u> surface. The size of pebbles and their arrangement on the fabric surface determine the type of

小卵石纹/卷曲

crepe fabric (crepe-de-chine, crepe-Georgette, and so on).

The crepe effect can be achieved either by the use of crepe yarns or a crepe weave, or sometimes by special process of finishing, i. e. embossing. The fabric is embossed with a metal roll, which has a raised relief engraving, under the conditions of high temperature and pressure. The surface of fabric changes from smooth into uneven or irregular.

Using the crepe yarns either in warp or weft gives a crepe fabric. Crepe yarns are tightly twisted. These yarns are composed each of two pairs of untwisted singles, where one pair is tightly twisted in the S direction and the other in the Z direction, and then both pairs are twisted around each other with a low twist. Crepe twist yarns of the fabric snarl and shrink during the wet finishing treatment throwing up pebbles wherever there is the least resistance. This method of crepe producing was widely used with natural silk.

Crepe weaves can be constructed on the basis of elementary weaves through removing the monotony of the fabric surfaces of these weaves by means of changing the arrangement of warp overlaps. There are no-general rules for the construction of crepe weaves, but many different methods are known.

4.2.1 Construction of crepe by drawing one weave over the other

In this method the starting point is to choose at least two weaves. One of the weaves is very often sateen. The warp repeat of this weave determines the number of shafts. Therefore, the type of shedding motion of the loom should be taken into consideration. It is common to choose the second weave with the same repeat as the first one. If the repeat of the second weave is not equal to the repeat of the first one, the repeat of the crepe weave is found as the least common multiple of the repeats of base weaves.

In the example shown in Fig. 4.4 the repeat of the sateen at (A) is equal to the repeat of the twill weave at (B). The crepe weave is constructed at (C). In constructing the crepe weave, all the shaded squares of diagrams at (A) and (B) are transferred to the weave diagram at (C). The constructed weave is characterized by non-uniform arrangement of the shaded squares. The shaded squares are concentrated on the warp threads 2 and 5 and weft threads 4 and 5. As a result the arrangement of warp overlaps becomes irregular. The fabric with such a weave has a pebbly surface.

(A)　　　　　　(B)　　　　　　(C)

Fig. 4.4　Construction of crepe weave

The base weaves for constructing a crepe weave by this method can be also the twill weaves with the opposite direction of twill lines.

4.2.2 Construction of crepe by arranging the warp overlaps in sateen order

在缎纹组织上加组织点构成绉组织

This crepe weave is constructed on the basis of sateen weave by adding the warp overlaps, i. e. the shaded squares on the weave diagram. Different groups containing two or more shaded squares can be added at one of the sides of each of shaded squares arranged in sateen order. The crepe weave constructed on the basis of sateen $\frac{7}{4}$ is shown in Fig. 4.5. Two shaded squares are added above and at the right side of each of seven shaded squares of the base sateen weave. Due to this the surface of new weave becomes uneven, and non-smooth.

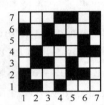

Fig. 4.5 Construction of crepe by rearranging warp(1)

4.2.3 Construction of crepe by rearranging warp

调整经纱排列构成绉组织

Crepe effect can be obtained by rearranging the warp threads of the base weave. Twill weaves can be chosen as basic ones. Then the order of warp threads of the base weave should be changed. The warp repeat of constructed weave is also changed very often, but the weft repeat remains the same. Many crepe weaves can be constructed on the same basis and all these crepe weaves can be produced on the number of shafts. It is very important in practice.

The arrangement is usually done by changing the position of the threads on the shafts, and then by constructing the weaving plan. Now it is necessary to arrange the warp threads on the shafts. No general rules for the arrangement of the warp threads can be given, because a variety of arrangements are possible. All depends on the skill of the designer which constructs several crepe weaves and chooses the best. Sometimes it is necessary, however, to produce samples of the fabric and study their appearance.

织小样

This method is illustrated in Fig. 4.6. The twill $\frac{2}{1}\frac{2}{3}$ at (D) is chosen as the base. The position of crosses on the lifting plan at (C) is similar to the position of shaded squares on the diagram of twill at (D).

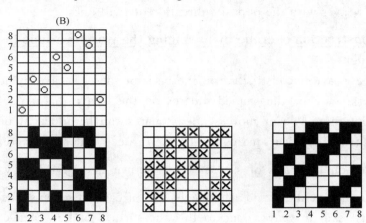

Fig. 4.6 Construction of crepe by rearranging warp(2)

The most important step is to arrange the circles on the draft. More often the curved draft is constructed. Now, two elements of the weaving plan are known and the third element, i.e. the weave, can be constructed at Fig. 4.5 (A). The warp overlaps on the crepe weave at Fig. 4.5 (A) are concentrated in four groups, making the surface of the fabric <u>irregular</u> and creating the <u>pebble</u> effect.

4.2.4 Construction of crepe by rearranging warp and weft

A crepe weave can be constructed by rearrangement of warp threads of the base weave, and then by rearrangement of the weft threads of constructed weave. A lot of crepe weaves can be obtained with the same base weave by changing the order of rearrangement of both the warp and weft threads. But all these weaves can be produced on the same number of shafts, which is equal to the weft repeat of base weave. This method is illustrated in Fig. 4.7.

The lifting plan at Fig. 4.7(A) is constructed according to the base weave. The draft at Fig. 4.7(B) is constructed on the number of shafts equal to the weft repeat of the base weave. The draft determines the arrangement of warp threads in the weave at Fig. 4.7(C).

The order of weft threads in the weave is shown at Fig. 4.7(D). In the next step of construction, the order can be changed. At Fig. 4.7(E) the other order of weft threads is shown. The weft threads at Fig. 4.7(F) are rearranged according to the order at Fig. 4.7(E) for creating the crepe effect.

If the designer has decided to start producing the fabric with this crepe weave, he must construct the weaving plan according to two known

UNIT I   General Knowledge and Simple Construction   65

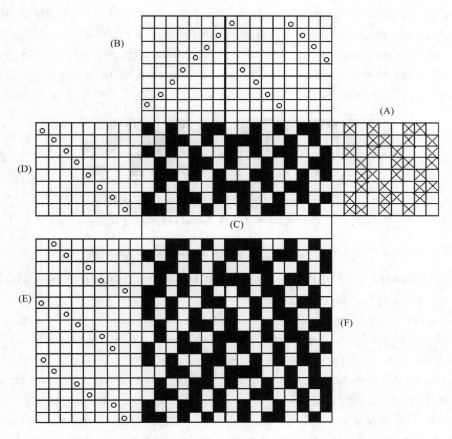

Fig. 4.7   Weaving plan of crepe weave

elements, i.e. the weave at Fig. 4.7(F) and the draft at Fig. 4.7(B).

4.2.5 Construction of crepe by placing the warp threads of one weave among the threads of the other weave

一种组织移到另一组织中间构成绉组织

In this method two base weaves should be chosen. It is better if the repeat of the second weave at (B) in Fig. 4.8 equals the repeat of the first weave at (A). The repeat should not be more than 5. It is desirable to have the number of warp overlaps equal to the number of weft overlaps. Then the ratio of the warp threads of the base weaves in the crepe weave is chosen. More often the ratio is 1 : 1. In this case the warp repeat of the crepe weave at (C) is twice of that of the base weave. The weft repeat of the crepe weave is equal to the weft repeat of the base weave. Now, the warp threads of the second weave are placed among the warp threads of the first weave and the warp threads of the first weave become the <u>odd threads</u> of crepe weave at (C), the warp threads of the second weave become the <u>even threads</u> of the crepe weave.

单号纱
双号纱

In this method the number of base weaves can be more than two and they can alternate in different ways. In all the cases the number of shafts on the draft equals the sum of shafts necessary for producing each weave separately. It is common to employ the divided draft for these crepe weaves. In this type of draft the shafts are divided into groups, the number of which equals the number of base weaves.

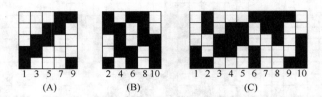

Fig. 4.8　Construction of crepe by combination of threads

The similar crepe weaves can be constructed by placing the weft threads.

4.2.6　Construction of crepe by placing the warp threads of one weave among the threads of its reverse

The crepe weaves constructed by this method can be produced on the tappet shedding motion with double tappets. On this motion the same tappets can be used for producing the weave or its reverse. The first step in constructing is to choose the base weave at (A) in Fig. 4.9. It is better when the number of warp overlaps is equal to that of weft overlaps. Then the reverse of the weave is constructed at (B) by replacing the warp overlaps with the weft ones and vice versa. To get the crepe weave at (C), all the even threads of the first weave are replaced by the even threads of the reversed weave. To demonstrate the development and for the sake of clearness, the base weave and its reverse are shown on separate drawings at (A) and (B), respectively.

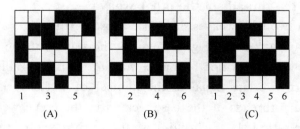

Fig. 4.9　Construction of crepe placing warp threads

旋转法构成绉组织

4.2.7　Construction of crepe by the method of rotation

In case of using this method the base weave should be chosen. Fundamental, derived, and crepe weaves with small repeat can be used in this method as base weaves. At (A) in Fig. 4.10 the base weave is shown. This weave is turned through 90° in a certain direction, for instance, clockwise as at (B). The weave at (B) is next turned one quarter way round to get the weave at

(C). Another quarter turn gives the weave shown at (D). Then all these weaves are transferred to the same drawing at (E) to make a crepe weave. The fabric of this weave can be produced on eight shafts with straight draft.

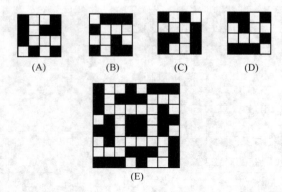

Fig. 4.10 Construction of crepe by method of rotain

A variation of this method consists in reversing the weaves at (B) and (D), or sometimes with another base weave by placing one weave over the other. A variety of different crepe weaves can be constructed by this method, but the designer should be skilled enough to determine in advance what effect the constructed crepe weave will have in ready fabric. In many cases, making fabric samples of the designed crepe weaves is advisable.

4.2.8 Construction of crepe by saving shaft method

省综设计法
构成绉组织

On the surface of crepe weave fabric, we don't want any regular patterns such as diagonal lines. In order to achieve this, very big repeat of the weave have to be selected. To avoid to many shaft, we use the saving shaft method practically.

The saving shaft method is processed as following steps:

(1) Determine the shafts, see Fig.4.11, 6 shafts is selected.

(2) Design the repeat. The warp repeat R_0 should be divided by the shaft selected and the weft repeat should be close to the warp repeat. Here, $R_0 = 6 \times 10 = 60$. $R_y = 40$.

(3) Determine the movement of each shaft. We need pay attention about:
- The floats of each end should be less than 3;
- The number of the intersection in each end should be close;
- The warp floats should be close to weft float of each ends.

(4) Draft plan

We often divided the warp repeat into groups, and rearrange the shaft sequence in each group. Some times, we need work for several times until the fabric's appearance is meet the requirement we need.

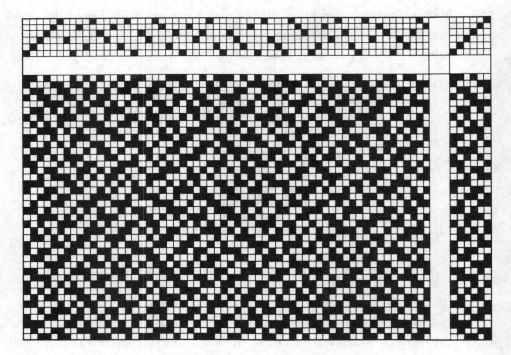

Fig. 4.11 Construction of crepe by saving shaft method

假纱罗组织

4.3 Mock leno weaves

These fabrics form an open structure with small holes or gaps similar to leno weave fabrics. These fabrics produce an imitation of leno effects and, due to this, the weaves of these fabrics are called the mock leno weaves.

窗帘
服装/内衣

These light and open fabrics are used for many articles, such as curtains, dress, underclothing, and so on.

The mock leno weave with the repeat of 8 is shown in Fig. 4.12. From the examination of the weave given in the figure two types of thread pairs can be found. The first type includes the weft threads 2 and 3, 6 and 7, and the warp threads of the same numbers. The threads of the second type are 4 and 5, 8 and 1 of the next repeat.

The weft threads 2 and 3 have the same order of interlacing. They float under the warp threads 1, 2, 3 and 4, and over the threads 5, 6, 7 and 8. No warp threads are placed between these weft threads. At the moment of beating-up, the weft thread 3 is pressed by the reed to the weft thread 2, forming no space between them. The same takes place while the weft thread 7 is inserted. The threads 7 and 6 are placed close to each other.

A pair of warp threads 2 and 3 interlace the weft in the same order. They

UNIT I General Knowledge and Simple Construction   69

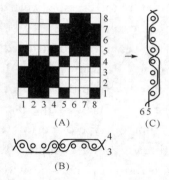

Fig. 4.12   Plan and sections of mock leno weave

should be drawn in the same <u>dent</u> of the reed, and this helps them to   筘齿
approach each other.

  The <u>adjacent</u> weft threads 4 and 5 are the threads of the second type. They   相邻的
interlace the warp in a different manner. The weft thread 4 passes over 1,
under 2 and 3, over 4 under 5, over 6 and 7, and under 8. The weft thread 5
interlaces the warp in reverse order, i.e. under 1, over 2 and 3, under 4,
over 5, under 6 and 7, and over 8. All the warp threads change their position
on the surface of the fabric, passing between weft threads 4 and 5 either
from the face to the <u>back side</u> of the fabric or vice versa. Thus, the warp   反面
threads prevent the weft threads 4 and 5 from coming together by forming an
open space in the fabric between these two threads. The other open space is
formed between weft threads 8 and 1 of the next repeat due to the same
causes.

  Examining the position of the adjacent warp threads 4 and 5, an opposite
order of interlacing of these warp threads in the weave can be seen. All the
weft threads pass between these warp threads and separate them. To increase
the gap, it is common to draw these threads in different dents of the reed.
The reed wire placed between these threads pushes them in opposite
directions. The open space is formed in the fabric at this point. The other
open space is formed between the warp threads 8 and 7 of the next repeat.

  The pairs of threads 1 and 2, 3 and 4, 5 and 6, 7 and 8 can approach each
other to some extent. Not 8 but only 4 threads of another system pass
between the threads of these pairs.

  Thus, the warp threads, as well as the weft threads, are grouped in fours
forming open spaces in the fabric.

  The formation of an open space in the warp direction can be seen at (B) in
Fig. 4.12. where the section through the weft threads 3 and 4 is given. The
warp threads 2 and 3, 6 and 7 are allowed to come together, while the warp

threads 4 and 5 are separated by the crossing weft threads and an open space is formed between these threads.

透孔   A variety of <u>perforated</u> weaves can be constructed on the basis of the lifting plan given at (C) in Fig. 4.13. The order of lifting of shafts 2 and 3 corresponds to that of plain weave and the order of lifting of shafts 1 and 4, to that of warp rib weave 2/2. The weaving plan is constructed with the

顺穿   <u>straight draft</u>. This is the simplest type of perforated weave. An open space is formed between weft threads 2 and 3, and between warp threads of the

缝隙   same numbers. To increase the <u>gap</u>, the reed wire is placed between these warp threads.

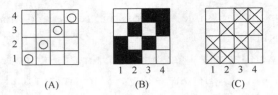

Fig. 4.13 Weaving plan of mock leno weave, $R=4$

    The weave of perforated fabric with repeat 6 can be constructed by using the lifting plan given in Fig. 4.14. Then the <u>pointed draft</u> is constructed at

山形穿法 (B) in Fig. 4.14. The weave at (C) is constructed from the known lifting plan at (A) and the draft at (B). The arrangement of weft threads of this weave is shown at (D). Then the weft threads are rearranged at (E) to get the position of circles similar to that on the draft at (B). The rearrangement gives a weave of perforated fabric with open spaces which are formed between threads 3 and 4, and 6 and 1 of the next repeat. For increasing the

筘齿   gap, it is common to draw three threads in one <u>dent</u>, i.e. 1, 2, 3 and 4, 5,

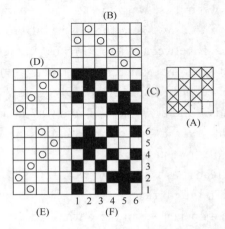

Fig. 4.14 Construction of weaving plan of mock leno weave

UNIT Ⅰ   General Knowledge and Simple Construction   71

6. Before producing this weave a weaving plan should be constructed from the known weave at (F) and the draft at (B).

This method can be employed for constructing weaves with repeat 8 as shown in Fig. 4.15, and with repeat 10 in Fig. 4.16. Both these weaves can be produced on four shafts.

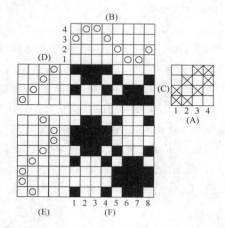

Fig. 4.15   Weaving plan of mock leno weave, $R=8$

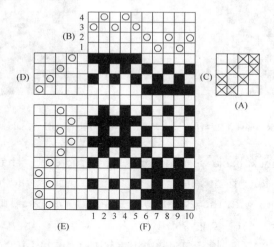

Fig. 4.16   Weaving plan of mock leno, $R=10$

Perforated weave is very often used in combination with other weaves, such as plain, sateen, and for making the figures in Jacquard designs.

4.4 Huckaback weaves

浮松组织

This weave contains, on one hand, a number of warp and weft threads with long floats which make the fabric soft and moisture absorbent, and, on the other hand, the plain weave threads which ensure the firmness of the

长浮线

坚实稳固

structure.

Huckaback weaves are used for bathroom towels, glass cloths, and for counterpanes.

These weaves are constructed on the basis of plain weave. The repeat of a huckaback weave usually contains an even number of threads and, due to this, includes the whole number of plain weave repeats. As a rule, the huckaback weaves are symmetrical about their diagonal axis. The huckaback weave shown in Fig. 4.17 contains 10 threads in the repeat. This weave has been constructed.

By adding two marks on each even warp thread in the left bottom quarter, and two marks on each odd warp thread in the right top quarter. Due to this, the diagonally opposite quarters are similar as in diaper designs. Two long warp floats and two weft ones are formed in the left bottom and right top quarters, the other areas are occupied with a plain weave. The long warp floats are on the face of the fabric, and the long weft floats are on the wrong side only.

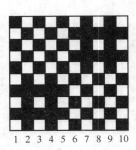
Fig. 4.17 Huckaback weave, $R=10$

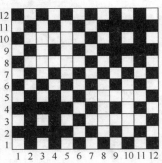
Fig. 4.18 Huckaback weave, $R=12$

A variety of huckaback weaves can be constructed by changing the repeat or the length of long floats. It is possible to construct the huckaback weave shown in Fig. 4.18 with the use of long floats in each quarter of repeat but, as a rule, if there are long warp floats in the left bottom and right top quarters, the long weft floats are constructed in the left top and right bottom quarters. The long warp floats on the threads 2, 4, 9 and 11 on the face of the weave are shown in Fig. 4.18. Four long weft floats are placed on the threads 3, 5, 8 and 10, where the weft thread in the place of float passes over five warp threads.

4.5 Honeycomb weaves

A group of weaves forms an embossed cell-like appearance of fabric. These so-called "cellular" fabrics are characterized by orderly distribution of hollows and ridges. The honeycomb weave is one of the most interesting weaves of this group. The surface of the weave looks like the honeycomb cells made of wax by the bees. The rough and loose structure of these fabrics makes them good absorbents of moisture. Due to this, these fabrics are widely used for bathroom towels and also for bedcovers, quilts, winter garments, and so on. To construct the weaving plan of honeycomb weave, the pointed draft should be drawn with the number of shafts from 4 to 8. The number of shafts equals 5 at (A) in Fig. 4.19. Then, the lifting plan is constructed at (B). The number of crosses, which is equal to the number of circles on the draft, is arranged similar to the position of circles.

凹的/隆起

浴巾/床罩/
被子/冬装

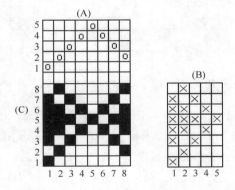

Fig. 4.19 Honeycomb weave, $R=8$

The ridges of honeycomb weave are formed by the longer floats of warp and weft. The warp ridge is formed by the first warp thread at (C), which is placed in a higher plane because there are only two weft overlaps on the first warp thread and this thread passes over six weft threads 1, 3, 4, 5, 6, 7. The eighth warp thread is also placed in a higher plane forming together with the first warp thread the second repeat (not shown in the figure) on the other warp ridge.

蜂巢组织

The weft ridge is formed by the first weft thread at (C), which passes over seven warp threads and is placed in a higher plane. The eighth weft thread is also placed in a higher plane, because it passes over six warp threads 1, 3, 4, 5, 6, 7 and there are only two warp overlaps 2 and 8. Thus, the ridges occur on the warp threads 1 and 8 and the weft threads 1 and 8. The point of intersection of the fifth warp thread and the fifth weft thread lies in the lowest plane and, as a result, a hollow is formed in the center part of the weave diagram.

The weaving plan of the honeycomb weave on four shafts is illustrated in Fig. 4.20. Two repeats of the draft are shown with 12 threads drawn in the healds. The ridges are formed by the threads 1, 7 and 12. The points of intersection of the warp threads 4 and 10 with the weft threads 4 and 10 are placed in the lowest plane, forming four hollows on this area of the fabric.

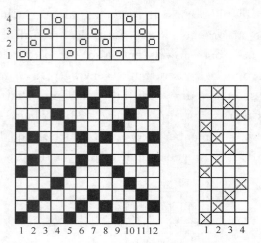

Fig. 4.20 Weaving plan of honeycomb weave, $R=6$

Besides the ordinary honeycomb weaves explained above, the Brighton honeycomb weaves are known, which are more complex in structure and produce the cellular formation with two sizes of cells on one side of the fabric. The wrong side of the fabric has a rough, indefinite appearance. These fabrics are produced in straight drafts. The repeat of draft should be a multiple of four, as well as the warp and weft repeats. Each repeat of this weave contains two large and two small cells formed by long warp and weft floats. The repeat of Brighton honeycomb weave shown in Fig. 4.21 contains 16 threads. Four hollows are formed due to interlacing of these threads.

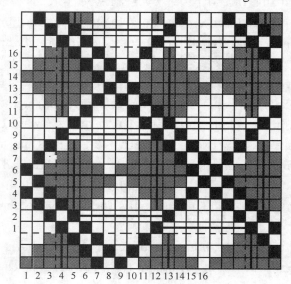

Fig. 4.21 Brighton honeycomb weave

UNIT Ⅰ   General Knowledge and Simple Construction   75

The honeycomb weaves are constructed by following steps.

(1) Constructe a diamond by single diagonal lines. See Fig. 4.22 (A). $\frac{1}{4}$ twill and $\frac{1}{5}$ twill are often used for the diamond base weave.   基础组织

(2) Fill in inside of the diamond with crosses leaving one row of squares between this space and the lines of the crosses blank. Now, a simple honeycomb weave is constructed. See Fig. 4.22 (B).

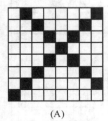 

(A)   (B)

Fig. 4.22   Construction of honeycomb weave

Larger honeycomb weaves are stitched with double diagonal lines to   固结
achieve a firmer structure without losing the honeycomb effect. See Fig. 4.23.

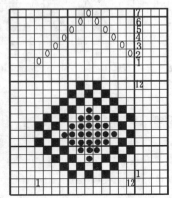

Fig. 4.23   Honeycomb are stitched with double diagonal lines

A variation of patterns can be created on the honeycomb principle. See Fig. 4.24.

Because of long floats in warp and weft, the sett and also the end use play an important part when deciding the type of variation.

The cell effect is more prominent with folded coarser yarns than with finer   明显/合股粗
yarns. The coarser fabrics, ideal for blankets, shrink considerably in the   线/收缩
finishing process.

Pointed draft is often used in honeycomb weaves.

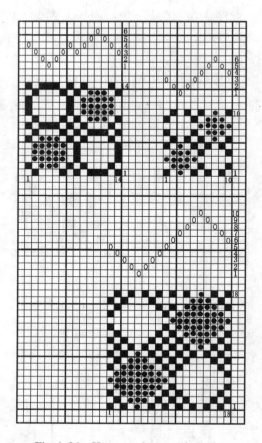

Fig. 4.24 Honeycomb weave variation

凸条组织

4.6 Bedford cord and piqué

圆形凸条/细沟槽

Both of these weaves are characterized by <u>rounded cords</u> with <u>fine sunken</u> lines between. The distinction between a Bedford cord and a piqué is that the former the cord runs along the length of the cloth. However, very few true piqués are made now.

The weave used is quite distinctive. See Fig. 4.25. The face of the cord is generally a plain weave. A twill is sometimes used instead, in which case the cloth was known at one time as "London cord". The rounded cord effect is achieved by pair of weft yarns floating across the back of the cord and being woven in to form the sunken lines.

纬重平

The Bedford weaves are constructed by combining a long float <u>weft rib</u> as base weave and a plain or a basic twill as face.

Example. Based on a $\dfrac{6}{6}$ weft rib and a plain for face of the cord draw a

Fig. 4.25 Bedford: section through warp

Bedford weave diagram (The face weave repeat should be factor of the base weave).

(1) Calculating the repeats:

R_0 = base weave repeat = 6 + 6 = 12,
R_y = weft repeat of base weave × weft repeat of face weave repeat
= 2 × 2 = 4.

(2) Drawing the outline of the repeat, and the base weave, weft rib. See Fig. 4.26(A).

(3) Filling the long weft floats with face weave. See Fig. 4.26(B).

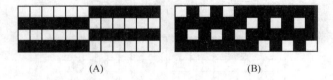

Fig. 4.26 Construction of Bedford

In production, we often modify the Bedford above, to put the two weft floats together in order to <u>accentuate</u> the effect of the cord. See previous Fig. 4.25.

加强

Bedford cords should be woven with a high sett in warp and weft.

The rib is accentuated through the plain weave <u>cutting ends</u> on either side of the cord, and by proper <u>denting</u> in separating them by the splits of the reed. The cutting ends are consumed at a different rate and may require a <u>separate beam</u> in production. See Fig. 4.27.

分条经纱

穿筘

另一织轴

It is also essential to select a strong yarn for the cutting ends, as in tests these ends tend to break first. Further emphasis can be given to the cord by using a harder twist for the weft. In finishing, the weft floats on the back will <u>shrink</u> more than the rest of the fabric, giving a greater roundness to the cord.

收缩

Adding extra thicker ends referred to as <u>wadding ends</u> will give the cord

填充纱

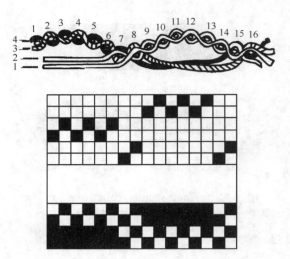

Fig. 4.27 Bedford with cutting ends

further stability and help to prevent it from becoming flattened. See Fig. 4.28.

Fig. 4.28 Bedford with wadding ends

The wadding ends are additional to the basic warp requirements and lie between the face of the cord and weft floats on the back of the cloth. Warp 4, 5, 14, 15 are wadding ends.

The number of wadding ends introduced depends on the density of the warp sett. A lower warp sett can accommodate more wadding ends than a higher one.

Figured Bedford cord

With this method, each unit of Bedford cord is divided by cutting ends running the length of the fabric and cutting picks stretching from selvedge to selvedge. Part of the weft is now not only floating on the back but also floating on the face of the cloth. See Fig. 4.29.

分条纬纱

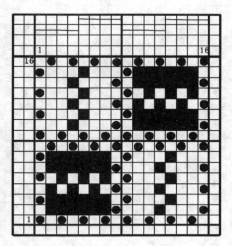

Fig. 4.29　Figured Bedford cord

Warp and weft faced cord units are arranged in vertical and horizontal direction. See Fig. 4.30.

Fig. 4.30　Warp and weft Bedford cord units

On point paper the width of the cord is marked first, then cutting ends and picks are entered in plain weave. The remaining spaces are then arranged according to the motif. Warp-faced cord sections are exchanged with sections where the weft is floating on the face of the fabric. Together they form the pattern.

花型

图案

In this example one square in the motif is equal to 6 cord ends and 2 picks. See Fig. 4.31.

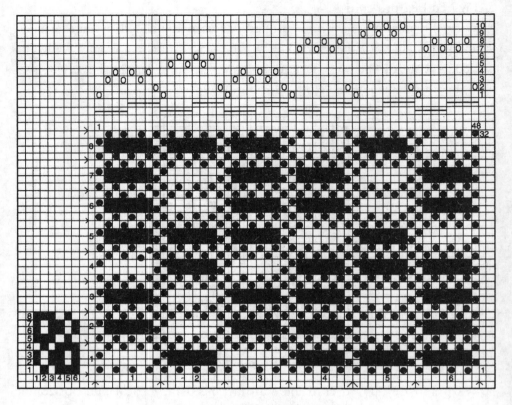

Fig. 4.31 Figured Bedford cord

This Bedford cord is arranged to give a waved effect; warp rib between the cords creates a cut to emphasize the cord effect. See Fig. 4.32.

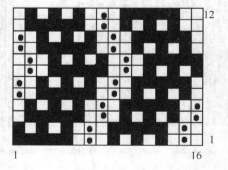

Fig. 4.32 Figured Bedford cord

This Bedford cord is arranged to give a diagonal effect; again warp rib between the cord creates a cut. See Fig. 4.33.

This warp stripe is a combination of Bedford cord, warp rib and plain weave, it requires a high sett. The total draft repeat consists of 240 ends.

The plain weave sections (θ) are woven from a separate beam with less

UNIT Ⅰ General Knowledge and Simple Construction   81

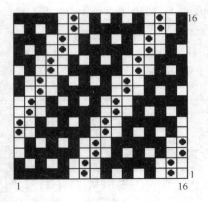

Fig. 4.33   Figured Bedford cord

tension than the main weave. After finishing, the plain weave <u>cockles up</u>,   皱起
creating a <u>pleated effect</u>. See Fig. 4.34.   褶裥

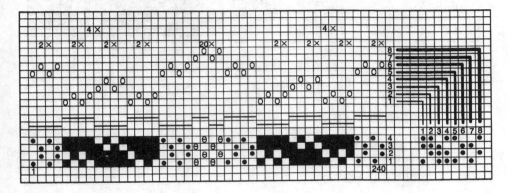

Fig. 4.34   Figured Bedford cord

4.7   Distorted weave effects

网目组织

4.7.1   Distorted weft effects (1)

With these structures very decorative fabrics can be produced with only four shafts.

Method of development.

Group (a): ends and picks interlace in plain weave.

Group (b): all ends float on the face,
           all picks float on the back.

Between succeeding groups two picks (c) interlace with all ends in groups (a) and (b) in plain weave.

The areas of tighter interlacing (a) build up in the normal way. But require more space than the areas of long floating threads (b). Which offer no resistance and allow picks (c) to move into this area. This causes a weft distortion and the formation of cell-like effects.

Simulated distorted effect that will appear on the face of the fabric. See Fig. 4.35, ■ ● ◓ = warp up.

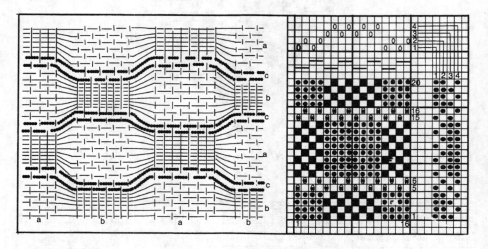

Fig. 4.35 Distorted weave effects

Example of sett.

Warp: approx. R 30 tex 2, 18 ends/1cm.

Weft: approx. R 30 tex 2 for groups (a) and (b).

Picks (c) can be of the same count in a different colour or of a coarser count to highlight the distortion.

It is advisable to weave these fabrics with a high warp tension and reduced weft tension.

4.7.2 Distorted weft effects (2)

The important point about these structures is that picks which are allowed to float on the face of the fabric move into the direction of least resistance. This causes the picks to diverge from their normal straight line and distort into zigzag lines.

All-over effects as well as horizontal stripes can be achieved with these weaves. End uses include apparel and curtain fabrics.

A predetermined number of ends and picks interlace in plain weave. At intervals ends (●) float alternately over the plain weave groups and interlace with the ground immediately before or after long floating picks have been inserted.

These picks do not interlace with the plain weave groups and are only held in place by interlacing with the long floating ends.

When the ground is beaten up the floating picks move into the direction of

UNIT Ⅰ General Knowledge and Simple Construction

least resistance and distort. The distortion is increased by employing <u>thicker</u> 粗号纱
<u>yarns</u> of <u>contrasting colours</u> for the floating picks. They should be inserted 对比色
with reduced tension.

Effect of the distorted weft is created in the fabric. See Fig. 4. 36.

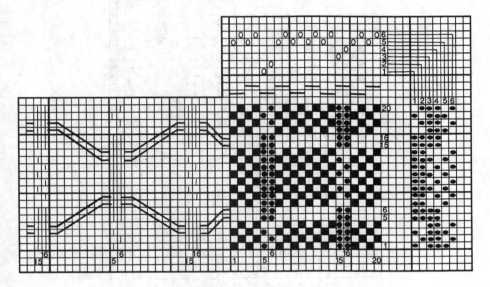

Fig. 4. 36 Distorted weave effects

4.7.3 <u>Distorted warp effect (1)</u>

经网目

In these constructions the ends are creating the distorted effect.

A number of weft floats are surrounded by plain weave ground. The weft
floats allow the end (⌷) to move into the area of <u>least resistance</u> causing a 阻力小
vertical zigzag line.

These ends are placed on a separate beam or roller, woven with less tension
and drawn in additionally to the ground ends into the reed.

Generally, thicker yarns of contrasting colours are employed to highlight
the effect.

The weave is arranged on a pointed draft and the shafts for the distorted
ends are located at the front.

To avoid long warp floats on the back of the fabric, ground ends are lifted
at certain intervals (■) to shorten these floats.

Effect of the distorted warp created in the fabric. See Fig. 4. 37.

4.7.4 Distorted warp effect (2)

In this construction the ends 4 and 10 are creating the distorted effect. The
first weft thread have a long floats from end 4 to 10 which offer no resistance
and allow the ends 4, 10 move closer. The weft thread 7 has a plain
interlacing with the ends from end 4 to 10 which offer a bigger resistance and

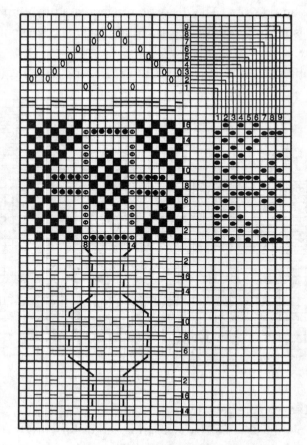

Fig. 4.37 Distorted weave effects

keep the ends 4 and 10 away. This causes a warp distortion and the formation of a vertical zigzag line. See Fig. 4.38.

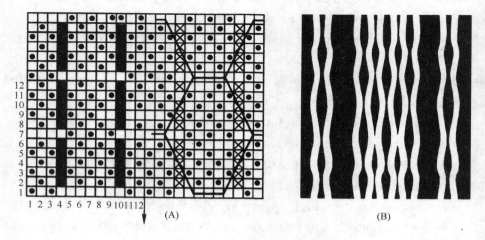

Fig. 4.38 Distorted weave effects

UNIT Ⅰ　General Knowledge and Simple Construction

4.7.5　Combined distorted warp and weft effects

经纬组合网目

To help in the development of this combination the ground weave has been drawn out without the distorting ends and picks. This gives also a clearer picture of the formation of the ground.

Arrows on the side of the diagram indicate the position where the desired effect have been inserted. Combined distorted warp and weft effect created in the fabric. See Fig. 4. 39.

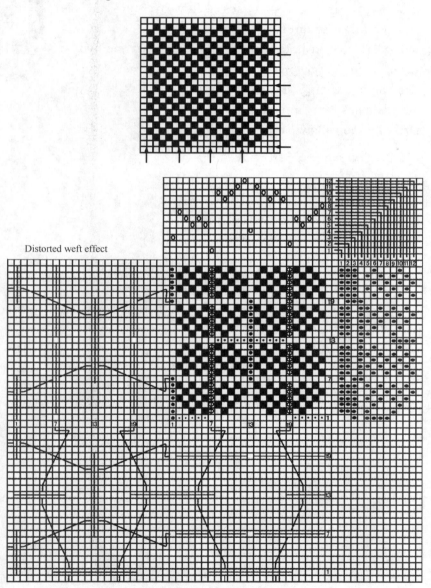

Fig. 4. 39　Combined distorted effects

HOMEWORKS

1. Constructing a crepe be rearranging warp the base weave is $\frac{3}{2}\frac{1}{2}\nearrow$, the sequence of the warp threads are 1, 3, 6, 2, 5, 4, 7, 8.
2. Constructing a crepe by the method of rotation with the base weave at Fig. 4.40, and each weave turns 90° in counterclockwise direction.
3. What are the advantages of the saving shaft method to produce crepe weave?

Fig. 4.40

4. Drawing a weaving plan of mock leno, $R_0 = R_y = 8$, and use the mark "O" to indicate the holes.
5. Drawing the following honeycomb weaves:

    (1) Base weave $\frac{1}{5}\nearrow$ $K_0 = K_y = 6$;

    (2) Base weave $\frac{1}{7}\nearrow$ $K_0 = K_y = 8$;

6. Drawing the following Bedford cords:

    (1) Base weave is 6/6 weft rib, face weave is plain;

    (2) Based on the previous weave, adding cutting ends on either side of the cord.

UNIT I   General Knowledge and Simple Construction   87

Chapter Five
Color and Weave Effects
色纱组织配合(配色模纹)

Various <u>coloured patterns</u> can be obtained in fabric by combining coloured yarns and weaves. That means, the pattern are not only depended on weaves, but also <u>colour sequence</u>.   花型花纹

色纱排列

The colour arrangements are called colour sequence. There are warp colour sequence and weft colour sequence. The number of a complete colour sequence threads is called <u>colour repeat</u>. There are warp colour repeat and weft colour repeat.   色纱循环

The <u>repeat of coloured pattern</u> equals the <u>least common multiple</u> (LCM) of the weave repeat and the colour repeat.   配色花纹循环/最小公倍数

The coloured patterns can be constructed by means of dividing the paper into four parts. See Fig. 5.1.

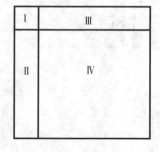

Fig. 5.1   Arrangement of coloured pattern construction

The arrangements are, I for weave, II for weft colour sequence, III for warp colour sequence, and IV for coloured pattern.

5.1 Construction of patterns from a given weave and colour repeats

根据已知组织图和色纱循环绘作配色模纹

For all two-tone colour and weave effects in this book only.
　　The dark coloured warp and weft floats are indicated: ■.
　　Light coloured warp and weft floats are left blank: □.

5.1.1 Method of development

The patterns are constructed as following steps.

(1) Determine the weave, warp colour sequence and weft colour sequence, e. g. Fig. 5. 2. The weave is $\frac{1}{1}$ plains; warp colour sequence and weft colour sequence are bath 1A, 1B, (A-light colour, B-dark colour). So the repeat of coloured pattern are two in warp and weft direction.

(2) Draw the weave, warp and weft colour sequence on their location, and fill the weave on colour pattern location by dot marks. In order to see the pattern clearly, more repeats are selected. See Fig. 5.2(A).

(3) According to the warp colour sequence, mark the warp floats (■) of the dark coloured ends. See Fig. 5. 2 (B). And according to weft colour sequence, mark the weft floats (■) of the dark coloured picks. See Fig. 5. 2 (C).

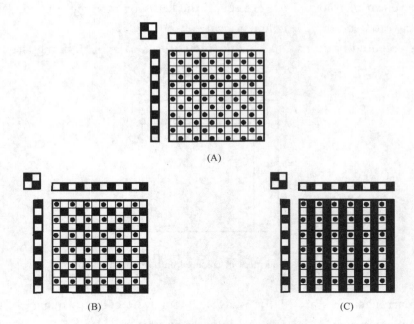

Fig. 5. 2 Construction of patterns

5.1.2 Varieties of weave and colour effects

5.1.2.1 Horizontal and vertical hairline effects (Fig. 5. 3)

UNIT I   General Knowledge and Simple Construction   89

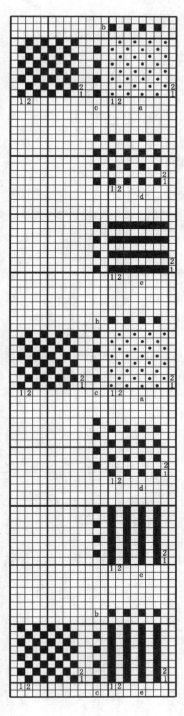

Horizontal effect
2 colours in warp and weft.

Warp and weft ■	1	
□	1	
		2

Method of development.
(a) Indicate the base weave very faintly.
(b) Indicate sequence of colours for warping order.
(c) Indicate sequence of colours for picking order.
(d) Mark the warp lifts (■) of the dark coloured ends.
(e) Mark the weft floats (■) of the dark coloured picks.

This diagram shows the colour and weave effects as seen on the face of the cloth.

Vertical effect
2 colours in warp and weft.

Warp ■	1	
□	1	
		2
Weft □	1	
■	1	
		2

Development as above.

For the vertical effect the colours for the picking order are reversed.

(e) This diagram shows the colour and weave effects as seen on the face of the cloth.

The vertical effect can also be achieved by reversing the base weave and using the same warp and weft colour sequence.

Warp and weft ■	1	
□	1	
		2

Fig. 5.3   Horizontal and vertical hairline effects

横竖条纹组合 5.1.2.2 Combined horizontal and vertical hairline effects (Fig.5.4)

竖条纹

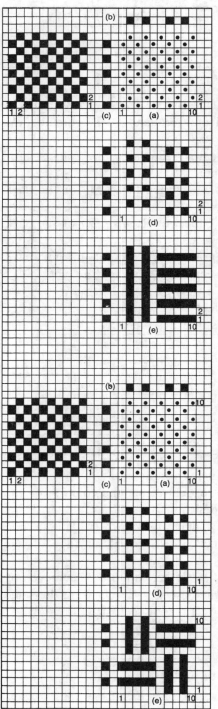

Stripe effect

2 colours in warp and weft.

Warp ☐	1	1	2	1	1	=	6
■	1	1	1	1		=	4
							10
Weft ■	1						
☐	1						
	2						

Development as shown on the previous page.

(e) This diagram shows the colour and weave effect as seen on the face of the cloth.

The "line change" occurs in this example when 2 light ends come together.

格型效果

Check effect

2 colours in warp and weft.

Warp and Weft ☐	1	1	2	1	1	=	6
■	1	1	1	1	·	=	4
							10

Development as shown on the previous page.

(e) This diagram shows the colour and weave effect as seen on the face of the cloth.

The "line change" occurs in this example when 2 light ends and 2 light picks come together.

Fig. 5.4 Combined horizontal and vertical hairline effects (1)

From now on the development stages are omitted and only the final colour and weave effect is shown. Each diagram however has originally been developed in stages in the same way as explained previously (Fig. 5.5).

过程/省略

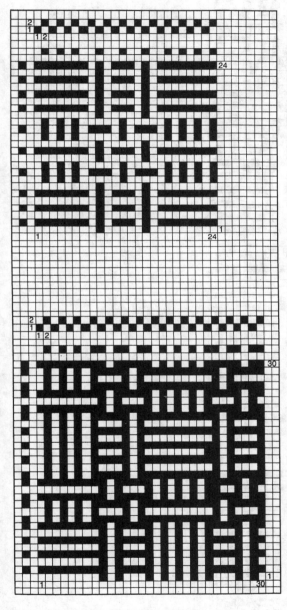

Fig. 5. 5 Combined horizontal and vertical hairline effects (2)

Figured effects

Base: plain weave.

Warp and Weft					
	3×	4×	3×		
☐	1	2	1	1	= 14
■	1	1	1	2	= 10
					24

Repeats:

12 weave = 24 ends/picks,

1 colour = 24 ends/picks.

The "line change" occurs in this example when 2 light ends and 2 light picks come together.

Warp and Weft					
	3×	3×	3×	3×	
☐	1	1	1	1	= 12
■	1	2	1	2	= 18
					30

Repeats:

15 weave = 30 ends/picks,

1 colour = 30 ends/picks.

The "line change" occurs in this example when 2 dark ends and 2 dark Picks come together.

图案效果　5.1.2.3　Figured colour and weave effects（Fig.5.6）

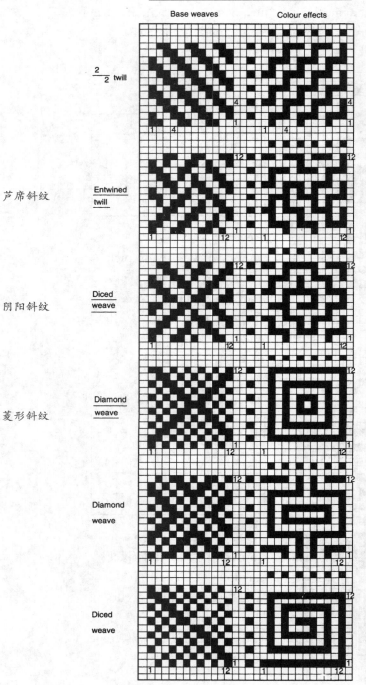

A great variety of effects can be created on one warp. The variations are achieved by changing the structure.

Warp and Weft ☐	1
■	1
	2

For this example the colour order in the weft has been changed.

Weft ■	1
☐	1
	2

Fig. 5. 6　Figured colour and weave effects

横竖条图案　5.1.2.4　Figured horizontal and vertical hairlines（Fig.5.7）

For this arrangement horizontal and vertical effects are entered to the pattern of a motif.

UNIT I General Knowledge and Simple Construction

Warp and Weft	■	1
	□	1
		2

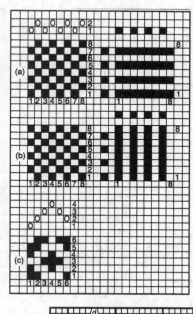

(1) Horizontal effect.
   Base weave (a), plain weave.
   1 group = 8 ends/8 picks.
(2) Vertical effect.
   For base weave (b) warp and weft floats of base weave (a) are exchanged.
   Warp and weft colour sequence remains the same. Therefore the "line change" from horizontal to vertical and vice versa can only be achieved by changing the weave. See also peg plan.
(3) Motif (c): each square represents 8 ends/8 picks. Motif repeat multiplied with either horizontal, or vertical effect equals total weave repeat (e):
   6 × 8 = 48 ends/48 picks.
(4) Full repeat (e) is marked out by following the motif:
   ■ = horizontal effect,
   □ = vertical effect.
(5) Draft (d): each symbol (0) of the condensed draft above the motif represents one group of two shafts containing 8 ends.

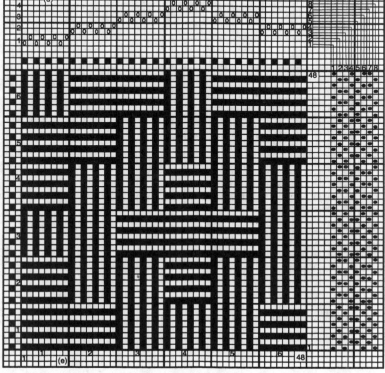

Fig. 5.7 Figured horizontal and vertical hairlines

粗细条图案   5.1.2.5   Single and double lines（Fig.5.8，Fig.5.9）

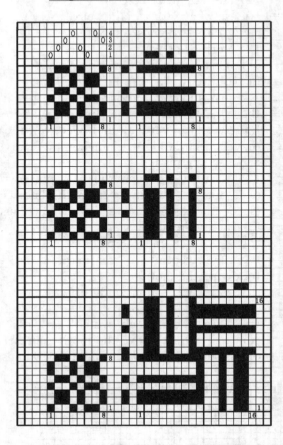

Horizontal effect

Warp and weft				
□	1	1	2	= 4
■	2	1	1	= 4
				8

Vertical effect

Warp	□	1	1	2	= 4
	■	2	1	1	= 4
					8
Weft	■	1	1	2	= 4
	□	2	1	1	= 4
					8

Check effect

Warp and weft						
□	1	1	2	2	1	1	= 8
■	2	1	2	1	2		= 8
							16

Repeats：
2 weave = 16 ends/picks，
1 colour = 16 ends/picks.
The effect will appear with a repeat that is almost always the lowest common multiple of the colour repeat and the weave repeat.

Step effect

Warp and weft				
□	1	1	1	1	= 4
■	1	2	1		= 4
					8

Repeats：
2 weave = 24 ends/picks，
3 colour = 24 ends/picks.

梯形纹

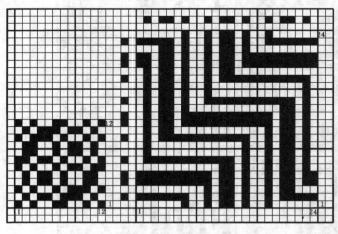

Fig. 5.8   Horizontal an vertical hairline effects（1）

UNIT I General Knowledge and Simple Construction

Warp and Weft ☐	1	1	1	1	=	4
■		1	2	1	=	4
						8

(1) Horizontal effect.
Base weave (a), 1 group = 8 ends/8 picks.
(2) Vertical effect.
For base weave (b) warp and weft floats of base weave (a) are exchanged. Warp and weft colour sequence remains the same.
(3) Motif (c): each square represents 8 ends/8 picks. Motif repeat multiplied with either horizontal or vertical effect equals total weave repeat (e):
6 × 8 = 48 ends/48 picks.
(4) Full repeat (e) is marked out by following the motif:
　　　　■ = horizontal effect,
　　　　☐ = vertical effect.
(5) Draft (d): each symbol (0) of the condensed draft above the motif represents one group of two shafts. containing 8 ends.

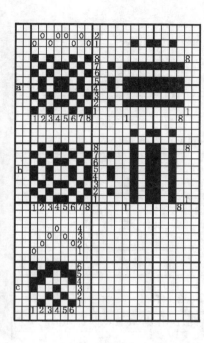

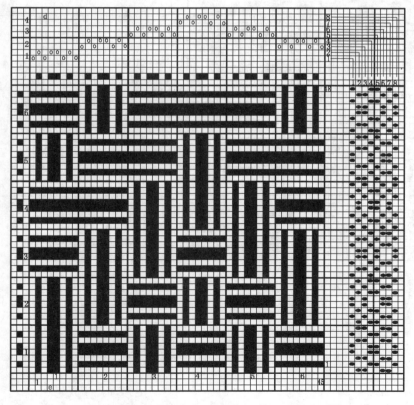

Fig. 5.9 **Figured horizontal and vertical hairlines (2)**

5.1.2.6 Double lines (Fig.5.10, Fig.5.11, Fig.5.12)

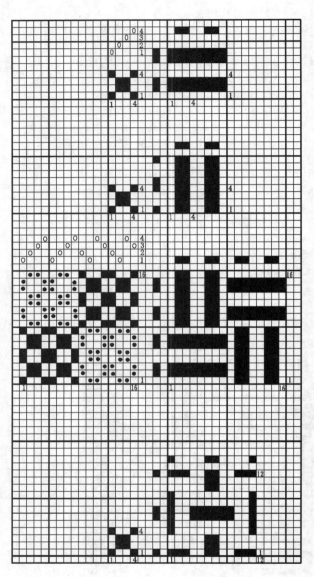

Horizontal effect

Warp and weft ☐	1	1	=	2
■		2	=	2
				4

Vertical effect

Warp ☐	1	1	=	2
■		2	=	2
				4
Weft ■	1	1	=	2
☐		2	=	2
				4

Check effect

Warp and weft ☐	1	1	=	2
■		2	=	2
				4

Repeats：
1 weave = 16 ends/picks,
4 colour = 16 ends/picks.

Warp and weft ■	1	1	=	2
☐		4	=	4
				6

Repeats：
3 weave = 12 ends/picks,
2 colour = 12 ends/picks.

Fig. 5.10 Double lines effects (1)

UNIT I General Knowledge and Simple Construction

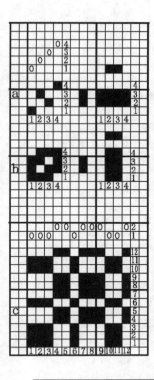

Warp and Weft ☐	1	1	=	2
■		2	=	2
				4

(1) Horizontal effect.
Base weave (a), weft faced, 1 group = 4 ends/4 picks.
(2) Vertical effect.
Base weave (b), warp faced, 1 group = 4 ends/4 picks.
(3) Motif (c): each square represents 4 ends/4 picks.
Motif repeat multiplied with either horizontal or vertical effect equals total weave repeat (e):
$\qquad$ 12 × 4 = 48 ends/48 picks.
(4) Full repeat (e) is marked out by following the motif:
$\qquad$ ■ = horizontal effect,
$\qquad$ ☐ = vertical effect.
(5) Draft (d): each symbol (0) of the condensed draft above the motif represents one group of four shafts containing 4 ends.

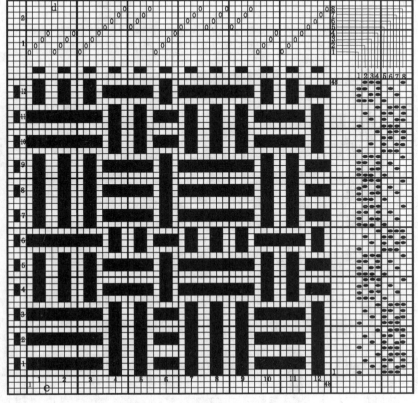

Fig. 5.11 Double lines effects (2)

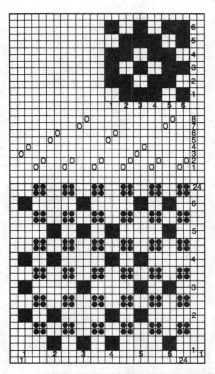

Motif.

This example shows how a base weave can be developed from a motif and then further progressed into a colour and weave effect creating nearly a Jacquard design with only 8 shafts.

A base weave with a hopsack foundation should be woven with a high sett.

For the development of the base weave the motif is extended by introducing double ends and double picks (●).

	Warp and weft										
□	2	4	2	2	2	2	2	4	2	=	24
■	2	2	2	4	2	4	2	2	2	=	24
											48

Repeats:
2 weave = 48 ends/48 picks,
1 colour = 48 ends/48 picks.

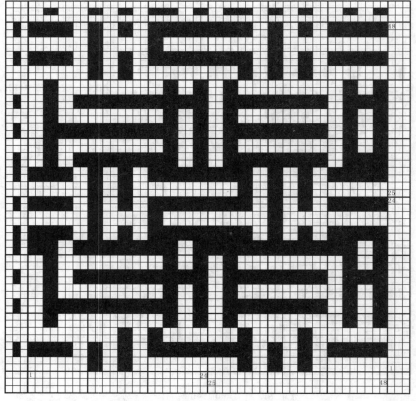

Fig. 5.12 Double lines effects (3)

5.1.2.7 Figured line effects (Fig.5.13, Fig.5.14)   不规则效果

All design on this page are based on $\frac{2}{2}$ twill.

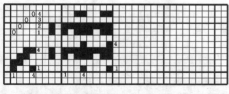

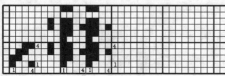

Horizontal effect

Warp and weft		
☐	2	
■	2	
	4	

Vertical effect

Warp		Weft	
■	2	☐	2
☐	2	■	2
	4		4

Stripe effect

Warp	☐	2	2	4	2	2	=	12
	■		2	2	2	2	=	8
						8		20
Weft	■	2						
	☐	2						
		4						

Check effect

Warp and weft							
☐	2	2	4	2	2	=	12
■	2	2	2	2		=	8
	4						20

Repeats：

5 weave = 20 ends/20 picks,
2 colour = 20 ends/20 picks.

To enlarge the design each section (＞) of the warping and picking order should be repeated to achieve the required size.

Check effect

Warp and weft					
☐	2	2	4	4	= 12
■	2	2	4	4	= 12
	4			24	

Repeats：

6 weave = 24 ends/24 picks,
1 colour = 24 ends/24 picks.

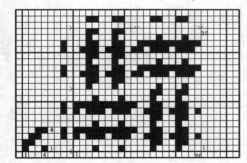

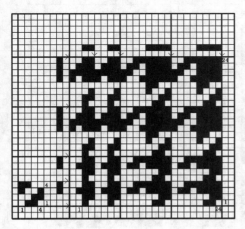

Fig. 5.13 Figured line effects (1), dogtooth effects   犬牙格

Warp				Weft			
□	1	1	= 2	■	1	1	= 2
■	2		= 2	□	2		= 2
			4				4

Base: diaper
Repeats:
1 weave = 20 ends/20 picks,
5 colour = 20 ends/20 picks.

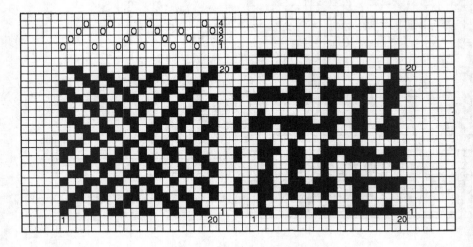

Warp and weft □	1	1	= 2
■	2		= 2
			4

Base: entwined twill
Repeats:
1 weave = 24 ends/24 picks,
6 colour = 24 ends/24 picks.

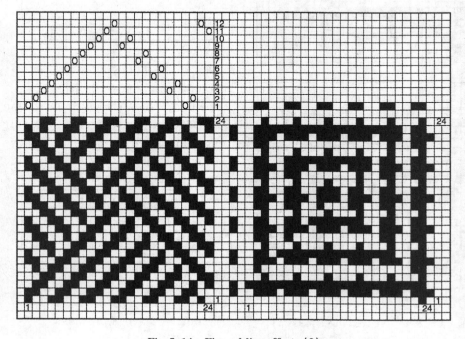

Fig. 5.14 Figured line effects (2)

UNIT I General Knowledge and Simple Construction   101

5.1.2.8 Stars and small figured effects (Fig. 5.15)   小花纹效果

Warp and weft □	1	1	=	2
■		2	=	2
				4

All examples on this page have the same colour sequence in warp and weft.
a = base weaves.
b = colour and weave effects.

The same star effect has been created with four different base weaves. Fig. 5.15(A).

In order to retain the same colour order for warp and weft and use it for more than one pattern, it may be found necessary to move the start of a weave repeat. For example the diced weave in the first example is not immediately recognizable as a diced weave. Fig. 5.15(B).

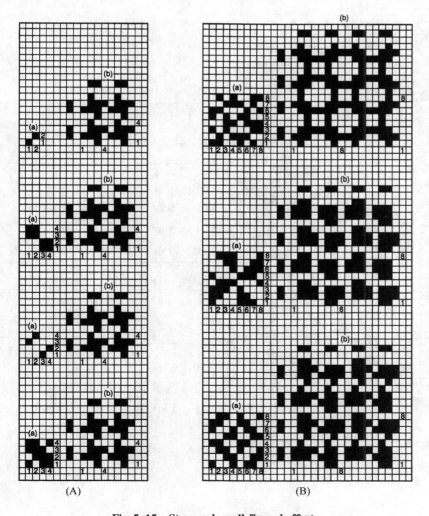

(A)   (B)

Fig. 5.15 Stars and small figured effects

小菱形效果　5.1.2.9 <u>Star and diamond effects</u>（Fig.5.16）

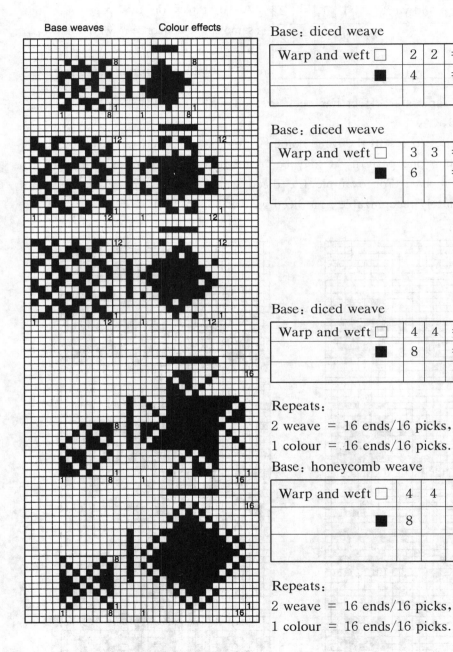

Fig. 5.16　Star and diamond effects

UNIT Ⅰ General Knowledge and Simple Construction

5.1.2.10 Shadow check effect (Fig.5.17)   阴影效果
Base: diced weave

Warp and weft □	3	4	3	2	2	3	4	3	=	24
■	2	3	4	6	4	3	2		=	24
										48

Repeats:
6 weave = 48 ends/48 picks,
1 colour = 48 ends/ 48 picks.

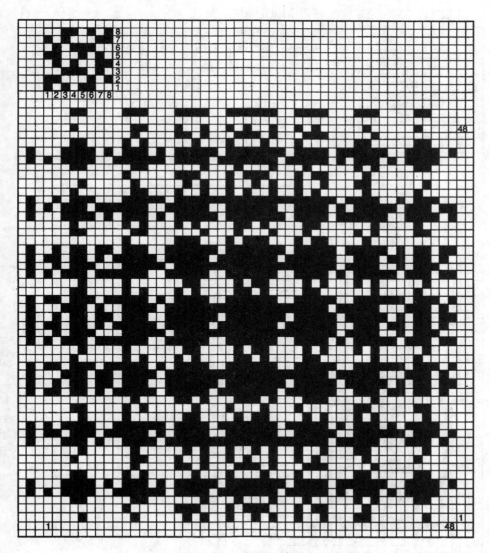

Fig. 5. 17 Shadow check effects

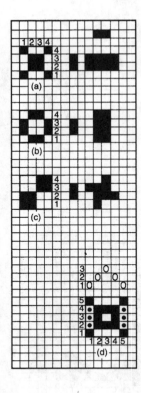

5.1.2.11 Figured check with three weave effects (Fig. 5.18)

(1) Horizontal effect:
base weave (a), split hopsack, 1 group = 4 ends/ 4picks.

(2) Vertical effect:
for base weave (b) warp and weft floats of base weave (a) are exchanged.

(3) Star effect:
base weave (c), regular hopsack.

(4) Motif (d):
each square represents 4 ends/4 picks.
Repeat: 5 × 4 = 20 ends/20 picks.

(5) Full repeat (e) is marked out by following:
□ = horizontal effect, weave (a),
● = vertical effect, weave (b),
■ = star effect, weave (c).

(6) Draft (f): each mark in the condensed draft above the motif represents one group of four shafts containing 4 ends.

Warp and weft □	1	1	=	2
■		2	=	2
				4

To enlarge the design each section of the warp and weft sequence has to be repeated to achieve the required size.

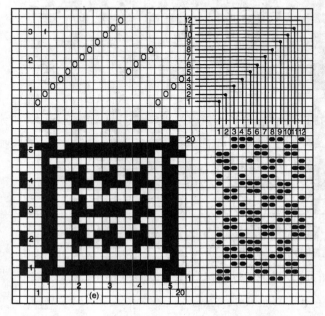

Fig. 5.18 Figured check with three weave effects

UNIT Ⅰ　General Knowledge and Simple Construction

5.2　Construction of weaves from a given pattern and colour repeats

已知色纱循环和配色花纹绘作组织图

When a fabric need to be <u>imitated</u>, that means the colour pattern and the colour repeat are given, the problem is to find and select the weave.

仿造

First of all, we should analyze the <u>intersections</u> of each warp threads and weft threads. Let's take the Fig. 5.19 for example.

交织点

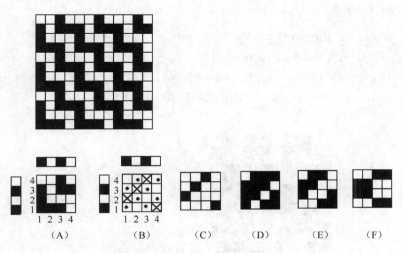

Fig. 5.19　Construction of weaves

The <u>odd</u> threads are dark colour, and the even threads are light colour. The colour repeat is 2 for warp as well as for weft threads. The arrangement of coloured warp threads is shown in the upper horizontal space, and that of weft threads in the left vertical space. Now we find the colour of all 16 squares the intersections of warp and weft threads. See Fig. 5.19 (A).

单数

The first square is the intersection of dark warp 1 and dark weft. So this square must be dark colour, no matter it is warp float or weft float. We put a mark dot "■", which indicates warp float or weft float. See Fig. 5.19 (B). The intersection of dark warp 1 and light weft 2 is dark colour. So this square must be warp float. We put a mark cross "⊠", which indicates warp overlap. The point of intersection of dark warp 1 and dark weft 3 must be dark colour whether it is warp float or not. So this square is marked by dot. The point of intersection of dark warp 1 and light weft 4 is light colour. So this square must be weft float. We do not need to put anything on it.

Thus, all 16 squares can be divided into three groups: dark, light, and squares marked by dots. These are 8 squares marked by dots. Each of these squares can be warp float or weft float at the wish of the designer. Many different weaves can be created. See Fig. 5.19 (C), (D), (E), (F).

5.3 Construction of weaves and colour repeats from a given pattern

A coloured pattern play a major role in aesthetics property of a fabric. It will be first determined usually. The problem is to choose a proper coloured threads sequence and a suitable weave for the pattern.

Following steps are suggested for the construction.

5.3.1 Presume the weft colour sequence

See Fig. 5.20(A), the weft threads 1, 3 have more dark colour intersections, so we select dark colour for weft threads 1, 3. Similarly, light colour are selected for weft threads 2, 4. See Fig. 5.20 (B).

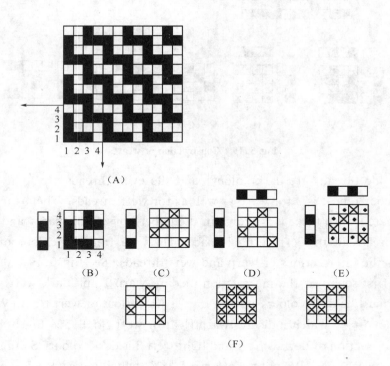

Fig. 5.20 Construction of weave and colour repeats

5.3.2 Surmise the inevitable warp floats

Surmise the warp float are according to the weft colour sequence we have selected previously. If the colour of the intersection is differ from the colour of weft thread. It must be a warp float. Based on this principle, the inevitable warp floats will be selected. See Fig. 5.20 (C).

UNIT Ⅰ General Knowledge and Simple Construction

5.3.3 Surmise the warp colour sequence

推测纱纱排列

After the inevitable warp floats are selected, check each warp float and make sure that all the inevitable warp floats in each end are the same colour. If there are more than one colour of the inevitable warp floats in one end, the presumption of the weft colour sequence is incorrect. It need to do it again. If there is only one colour of the inevitable warp floats, the previous presumption is correct. According to the colour of the warp floats, the warp colour sequence can be easily decided. See Fig. 5. 20 (D).

Based on colour repeats, we just got, and the pattern given, analyzing each intersections, we get Fig. 5. 20 (E) and various weaves at Fig. 5. 20 (F).

HOMEWORKS

1. Drawing the following coloured pattern from the given weave and colour repeat. (If it is necessary, you can select more repeat to show the pattern clearly).

(1) Base weave plain:

Warp and weft colour repeat 1 dark (1D hereafter), 1 light (1Lhereafter), (pattern repeat 8).

(2) Base weave:

Warp and weft colour repeat, 1D, 1L.

(3) Base weave:

Warp and weft colour repeat are:

☐	1	1	2	1	1
■		1	1	1	1

(4) Base weave:

Warp and weft colour repeat, 1L 2D 1L, (Pattern repeat 8).

(5) Base weave:

Warp and weft colour repeat: 4L 4D, (Pattern repeat 16).

(6) Base weave:

Warp and weft colour repeat: 1D 1L, (Pattern repeat 12).

(7) Construct the following weaves from the given pattern or with color repeat.

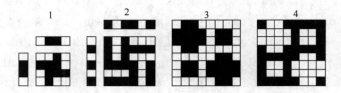

UNIT II

Compound Structure
复杂组织

Fundamental and combined weaves are considered to be simple, though there is a great variety of types and constructions. In these weaves, only one system of warp threads is interlaced with one system of weft threads at right angles. Due to this, the methods of construction of these weaves and production of fabrics of such weaves at textile mills are rather simple. 一个系统

Compound weave fabrics are of a specific structure, therefore special methods and mechanisms should be used in their production.

More than one system of warp and weft threads are used in these weaves. Very often the threads of different systems are arranged in different planes, forming two or more layers. If the tension in weaving or crimp of the warp of additional system differs from those of the system of the ground thread, a separate weaver's beam is necessary. If the systems of weft threads differ in yarn count type of fibers or colour, the loom should be equipped with a multi-shuttle mechanism. 层

多梭箱

In some kinds of compound weave fabrics, special systems of warp or weft threads are used and the threads of these systems after being cut in the process of weaving or finishing form a pile on the surface of the fabric. 绒头

There are compound weaves where the warp threads of additional system are not parallel to the ground threads, and have different position in the fabric forming an open structure.

The thickness and weight of one of the classes of compound weave fabrics are rather great because of additional systems of warp and weft threads. 厚度

For producing the compound weave fabrics, it is necessary to have a dobby on the loom and special design motions, such as slay motion, multi-shuttle mechanism, let-off motion, and mechanism for cutting the pile. They are necessary for producing particular effects on the fabrics. 多臂
筘型

Chapter Six
Backed Weaves
二重组织

The principle of backing a cloth with a second series of either weft or warp threads is to add extra weight and warmth without interfering with the smooth surface of the fabric. The end uses of backed cloths range from apparel to furnishing.

6.1 Warp backed weaves

These are weaves, which have two systems of warp and one system of weft. The face weave is formed by interlacing face warp and weft. The back weave is formed by interlacing back warp and weft. The system of weft plays an important role due to interlacing with both systems of warp. The weft threads are raised above the face warp and lowered the back warp.

In order to construct a perfect backed warp weaves, the following points need to be paid for attention.

6.1.1 Selection of face weave and backed weave

The face weave can be the same as the back weave, and differ from the back weave, but the face weave should be warp-faced weave such as $\frac{3}{1}$ twill, back weave should be weft-faced weave such as $\frac{1}{3}$ twill.

6.1.2 Selection of the starting point in construction of a back weave

In order to get a better appearance of the fabric, the backing stitches (back warp floats) should be hidden between floats on the face. So the warp floats of the back weave should be in the middle of the adjacent warp floats of the face weave. The setting of the face warp should be dense enough to prevent the binding marks of the back warp showing through.

6.1.3 Determine the arrangement of the face and back warp

The threads of the back warp can be arranged either alternately or in the proportion of two face threads to one back thread, i. e, $m:n = 1:1$ or $m:n = 2:1$.

6.1.4 Calculate the "new repeat"

$$R_0 = LCM \text{ of } \left(\frac{LCM \text{ of } R_m \& m}{m} \text{ and } \frac{LCM \text{ of } R_n \& n}{n}\right) \times (m+n)$$

$$R_y = LCM\ R_m \& R_n$$

Here: LCM—least common multiple;

R_0—the warp repeat of the backed weave;

R_y—the weft repeat of the backed weave;

R_m—face weave warp repeat;

R_n—back weave warp repeat;

m, n—the warp arrangement of face weave and back weave.

E. g. A backed weave:

face weave: $\frac{3}{1}\nearrow$;

back weave: $\frac{1}{3}\nearrow$;

warp ration: $1:1$;

$$R_0 = LCM \text{ of } \left(\frac{4\&1}{1} \text{ and } \frac{4\&1}{1}\right)(1+1) = 4 \times 2 = 8;$$

$$R_y = LCM\ R_m \& R_n = 4.$$

Construction of a backed warp weave diagram.

(1) Select: the face weave $\frac{3}{1}$ Z twill, the back weave $\frac{1}{3}$ Z twill. See Fig. 6.1. (A), (B). Arrangement $1:1$.

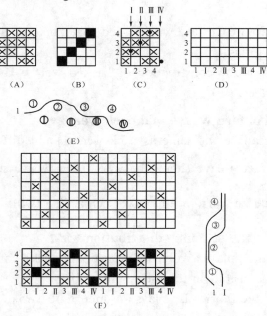

Fig. 6.1 Construction of backed warp weave

(2) Determine the starting point which must meet the previous point (2), i. e. stitching float should be covered by the face warp float. See Fig. 6.1 (C).

The Arabic numbers indicate the face warp ends.

The Roman numbers indicate the back warp ends.

The dots indicate the back warp threads over the weft threads.

(3) Calculate the repeat and outline the repeat.

Here: $R_0 = 8$ $R_y = 4$. See Fig. 6.1 (D).

(4) Transferring the weaves.

See Fig. 6.1 (E).

We usually drawing-in the face ends into the front shafts. The same group ends should be drawn into one dent of the reed, in order to cover the stitching float properly. When the tension of the two sets of warp are different, a second beam is required.

6.2 Weft backed weaves

These are weaves which have two systems of weft and one system of warp. The face weave is formed by interlacing warp and face weft. The back weave is formed by interlacing warp and back weft.

Weft faced structures are more suitable for this method as the backing stitches can be hidden between floats on the face. The beating up of the face picks close together forces the stitches to be covered so that they are not visible on the face, and the face picks are not visible on the back of the fabric.

The backed weft weaves are widely used for blankets, thick woollens and some industrial fabrics.

The principles of the construction of backed weft weaves are similar to backed warp weaves.

6.2.1 Selection of face weave and back weave

The face weave can be the same as back weave, and differ from the back weave, but the face weave should be weft-faced weave such as $\frac{1}{3}$ twill, back weave should be warp-faced weave such as $\frac{3}{1}$ twill.

6.2.2 Determine the stitching distribution

The correct stitching plays a very important part in the construction. On no account must it be visible on the face of the fabric.

The density of the fabric also plays an important part in achieving perfect stitching.

UNIT Ⅱ　Compound Structure

6.2.3　Determine the arrangement of the face and back weft

表、里纬纱排列

The threads of the back weft can be arranged either alternately or in the proportion of two face threads to one back thread, i. e., $m:n = 1:1$ or $m:n = 2:1$.

6.2.4　Calculate the new repeat

The calculation of the new repeat is similar to backed warp weaves. The difference is replacing the warp by weft.

Construction of a backed weft weave diagram is similar to backed warp weaves. We learnt previously. The same example are shown bellow.

Example 1. See Fig. 6. 2.

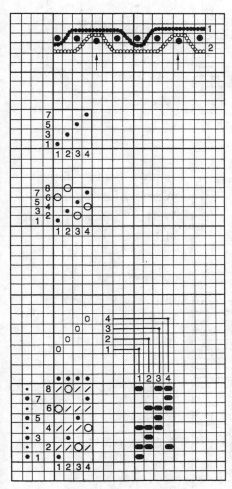

Fig. 6. 2　Construction of weft backed weave (1)

Cross-section:
face weave = pick 1,
back weave = pick 2.
Arrows = stitches.
Development of structure.

横截面图

接结处

Face weave: $\frac{1}{3}$ Z twill.

Back weave: $\frac{3}{1}$ Z twill.

Stitching is marked with circles (lowering of ends).
Horizontal lines between face picks represent back picks 2, 4, 6, 8. The back picks are stitched between weft floats of the face picks.
The stitching should be distributed in rotation over every end to avoid different tension on individual ends.
The correct stitching plays a very important part in the construction. On no account must it be visible on the face of the fabric.
Transferring the weaves.
Warp: solid. Weft: 1 face-1 back.
Repeat: 4 ends/8 picks.
Face weave on odd picks.
Back weave with stitching on even picks.
Lifters/: ends over back picks.
Circles are cancelled lifters (lowering of ends).
Reversible weave: identical interlacing on both sides of the fabric.
Completed structure with draft and lifting plan.
■ ╱ These symbols represent warp up.

Example 2. See Fig. 6.3.

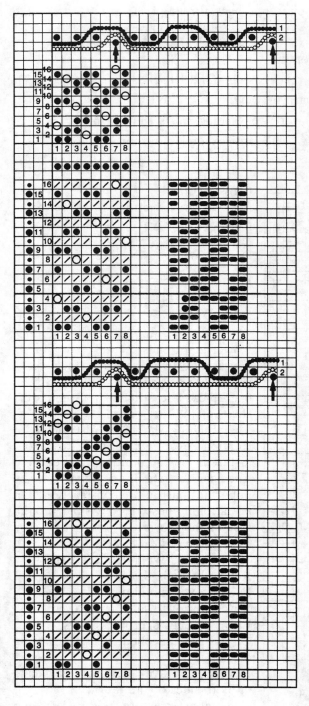

Fig. 6.3 Construction of weft backed weave (2)

Weft backed

In very dense warp setts the stitching points of the backing weft may be extended over two or more repeats of the face weave.

Cross-section:
face weave = pick 1,
back weave = pick 2,
arrows = stitches.

Face weave: $\frac{2}{2}$ Z twill, repeated twice.

Back weave: $\frac{7}{1}$ twill.

Repeat: 8 ends/16 picks.
Warp: solid.
Weft: 1 face – 1 back pick.

Cross-section:
face weave = pick 1,
back weave = pick 2,
Arrows = stitches.

Face weave: $\frac{2\ 1}{2\ 3}$ Z twill.

Back weave: $\frac{7}{1}$ Z twill.

Repeat: 8 ends/16 picks.
warp: solid.
Weft: 1 face – 1 back pick.

UNIT Ⅱ Compound Structure   115

Example 3. See Fig. 6.4.
Weft backed
Examples of reversible weaves. The heavier yarn count in weft covers the finer warp on both sides of the fabric entirely. These fabrics generally receive a milling and raising finish.

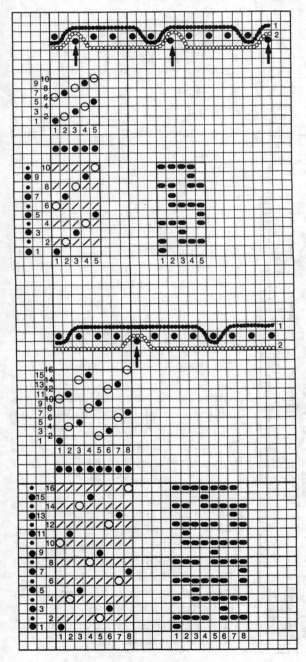

Cross-section:
face weave = pick 1,
back weave = pick 2,
arrows = stitches.

Face weave: $\dfrac{1}{4}$ sateen.

Back weave: $\dfrac{4}{1}$ satin.

Repeat: 5 ends/10 picks.
Warp: solid.
Weft: 1 face - 1 back pick.

Cross-section:
face weave = pick 1,
back weave = pick 2,
arrows = stitches.

Face weave: $\dfrac{1}{7}$ sateen.

Back weave: $\dfrac{7}{1}$ satin.

Repeat: 8 ends/16 picks.
Warp: solid.
Weft: 1 face - 1 back pick.
Identical interlacing on both sides of the fabrics.

Fig. 6.4 Construction of backed weft weaves (3)

Example 4. See Fig. 6.5.
Weft backed

Development of patterns with two effects. The cross-section shows the interchanging of face and back picks.

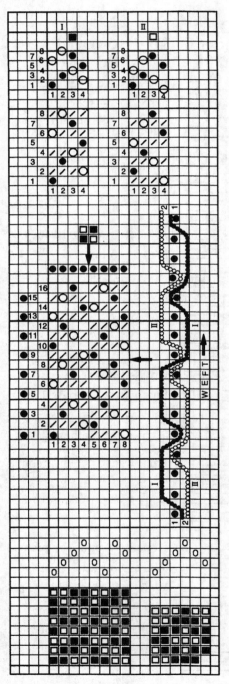

Effect I
Face weave: $\frac{1}{3}$ broken twill (pick 1, 3, 5, 7).
Back weave: $\frac{3}{1}$ broken twill (pick 2, 4, 6, 8).
Effect II
Face weave: $\frac{1}{3}$ broken twill (pick 2, 4, 6, 8).
Back weave: $\frac{3}{1}$ broken twill (pick 1, 3, 5, 7).
Check-pattern.
Motif
To enlarge the design each section of the structure can be repeated to achieve the required size.
Repeat of one section: 4 ends/8 picks.
Warp: solid.
Weft: 1 face - 1 back pick.
This example can be woven with 8 shafts.
When developing patterns, it is important to plan the weaves to cut with each other at the point of interchange between face and back. This assures clarity of design.

Motifs
Each square represents 4 ends/8 picks.
Each symbol of the condensed draft above the motif represents one group of four shafts.
These examples can be woven with 16 shafts.
These motifs can be developed with the same weaves as above.

Fig. 6.5 Construction of patterns of backed weaves

The proper set and yarn linear density are also important in achieving perfect appearance.   线密度

The warp density should be considerable lower, and the weft density should be higher due to the fabric effect depending the weft.

The count for the backing pick can be softer in twist, but should not be heavier than the face yarn, especially on a 1 face-1 back ratio.

The warp yarn should be stronger considerably due to its bearing the beating force during weaving process.   承受 打纬

HOMEWORKS

1. Construct backed warp weaves.

    (1) Face weave $\frac{3}{1}\nearrow$, back weave $\frac{1}{3}\nearrow$, arrangement of face warp and back warp $m:n = 1:1$.

    (2) Face weave $\frac{8}{5}$ satin, back weave $\frac{1}{3}\nearrow$, $m:n = 1:1$.

2. Construct backed weft weaves and their cross-section diagrams.

    (1) Face weave $\frac{1}{3}$ broken twill, back weave $\frac{3}{1}$ broken twill, $m:n = 1:1$.

    (2) Face weave $\frac{2}{2}\nearrow$, back weave $\frac{3}{1}\nearrow$, $m:n = 1:1$.

Chapter Seven
Multi-ply Fabrics
多层织物

7.1 Double fabrics

7.1.1 The concept of double fabrics

Double fabric consists of two layers which are woven one above the other. This fabric contains as a minimum two systems of warps, face and back, and two systems of weft. This fabric can also be called two-ply fabric. The upper layer is formed by interlacing the face warp threads with the face weft threads, and the lower layer by interlacing the back warp threads with the back weft threads.

Double fabrics are very useful that the single fabrics can not replace them. For instance:

(1) A tubular fabric can be formed by connecting both weaves at the edges. These fabrics used for ribbons, covering cylindrical objects, medical purposes, fire hose, etc. Various dimensions can be produced.

(2) A double width cloth can be made in conventional looms by connecting one edge of the weaves.

(3) Stitching the double fabrics to form a thicker and heavier fabrics.

(4) The different materials can be used in each layer to reduce the production cost.

7.1.2 The principles of double fabrics construction

The principles of double-weave construction are as following. See Fig. 7.1.

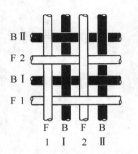

Fig. 7.1　Double weave

Fig. 7.1 shows a double fabric which face and back are all plain weave.

Here, the Arabic 1, 2 represent face ends and picks, and the Roman Ⅰ, Ⅱ represent back ends and picks.

The arrangement of face and back is 1 : 1, i.e., one face end, one back end; one face pick, one back pick.

For making double fabric, the picks are in sequence. When we fill the back pick, all the face ends must be up.

Suppose the face ends are drawn in shaft 1, shaft 2, the back ends are draw in shaft 3, shaft 4. The processes of the construction double fabric are described as following. See Fig. 7. 2.

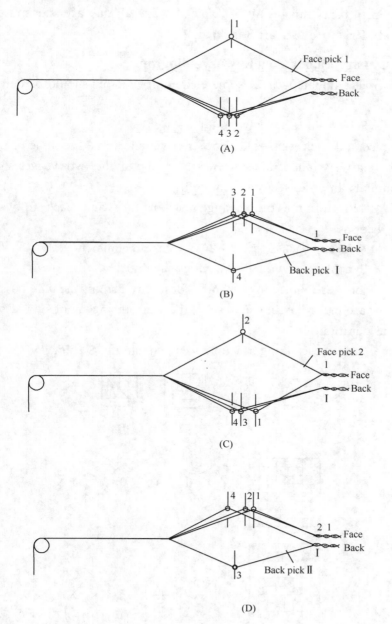

Fig. 7. 2 Construction of double fabric

(A) First pick: this pick is face pick 1, and all the back ends must be down. The shaft 1 is up.

(B) Second pick: this pick is back pick Ⅰ, and all the face ends must be up. The shaft 3 is up, and shaft 4 is down.

(C) Third pick: this pick is face pick 2, and all the back ends must be down. The shaft 2 is up.

(D) Forth pick: this pick is back pick Ⅱ, and all the face ends must be up. The shaft 4 is up, and shaft 3 is down.

7.1.3 Construction of double weave diagram

To construct a double weave, some elementary factors should be determined first.

7.1.3.1 Select the face weave and back weave

The face and back weaves are not strict which can be the same or not, due to their relatively independence. But the crimp of the two weaves should be quite similar, in order to weave smoothly

7.1.3.2 Determine the arrangement of face and back in warp and weft directions

The ratio face : back 1∶1, 2∶2, 2∶1 are recommended.

7.1.3.3 Synchronies with the weaving machine

If the looms are multi-box in both sides, any arrangement is possible. If the looms are multi-box in single side, the weft arrangement should be taken into consideration.

Example for constructing a double weave diagram. See Fig. 7.3.

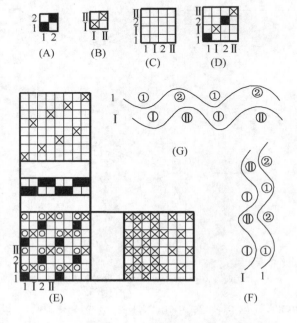

Fig. 7.3 Construction of double weave diagram

(1) Select face and back weaves, and draw them. Suppose the plain weaves are selected for both face and back weaves. See Fig. 7.3 (A), (B).

(2) Determine the arrangement $m:n$. Here $m:n=1:1$

(3) Calculate the double weave repeat.
The formula is

$$R_0 = LCM \text{ of } \left(\frac{LCM \ R_m \ \& \ m}{m} \text{ and } \frac{LCM \ R_n \ \& \ n}{n} \right) \times (m+n).$$

Here: R_m—face weave warp repeat;

R_n—back weave warp repeat;

$m:n$—the arrangement of the face and back in warp way.

R_y can be calculate by the similar formula.

In this example, $R_0 = 4$, $R_y = 4$, $m = 1$, $n = 1$.

(4) Draw the outline of the repeat, and indicate the number of each space. See Fig. 7.3 (C).

(5) Transferring the weaves. Transfer the face weave and back weave to the new weave. See Fig. 7.3 (D). 移

(6) Lifters: all face ends are lifted over back picks. The mark ⊡ represents the lifters. See Fig. 7.3 (E). 加表经提综符号

Cross-sections along the warp and weft directions are shown in Fig. 7.3 (F), (G).

The face ends are usually drawn into shafts in front. 前面的综框

7.2 Stitching double fabrics

接结双层

Two layers of the double cloth can be stitched together to form one of the principal features.

The stitching of the back and face fabrics of the double fabric can be effected in five ways. In the first three methods the threads from layers are used for stitching and they are also called self stitching. In the other two methods extra threads either warp or weft are introduced, which lie between the face and the back fabrics, stitching them and they are called center stitching. 实行

接结纱

The first method is to stitch from back to face, carried out by raising the back warp above the face weft. 下接上

The second method is to stitch from face to back, carried out by lowing the face warp below the back weft. 上接下

The third method, i. e. the combined stitching, allows face warp to stitch down and the back warp to stitch up. The warp of each fabric is included in 上、下对接

接结经

the shed of the other fabric.

The fourth method is called the stitching with an <u>extra warp</u>. It occurs when the face and back fabrics are stitched together by extra warp and there is no interlacing of the threads of the face fabric with those of the back fabric. Three systems of warp and two systems of weft are used in this case. It is necessary to distinguish the extra stitching warp from the extra stuffer warp, the threads of which can also lie between the face and back fabrics without interlacing with the weft threads.

接结纬

The fifth method is called the stitching with an <u>extra weft</u>. In this case, the face and back fabrics are stitched together by extra weft which binds the face and back warps. The fabrics are held together only by extra weft threads. For increasing the mass of the fabrics, stuffing weft threads can be introduced between the fabrics. When the extra weft threads differ in count or type of fibres, the loom should be equipped with a muti-shuttle mechanism.

7.2.1 Back cloth stitched to the face cloth (Fig. 7.4, Fig. 7.5, Fig. 7.6)

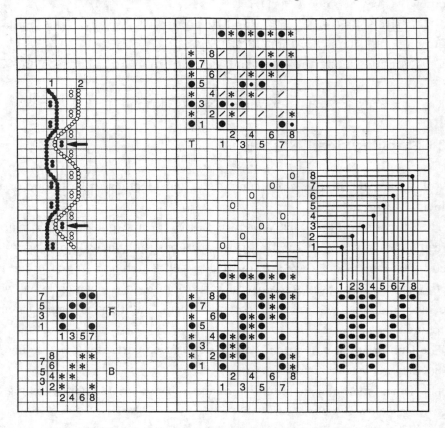

Fig. 7.4 Construction of stitching double weave (1) (Method Ⅰ)

UNIT Ⅱ   Compound Structure   123

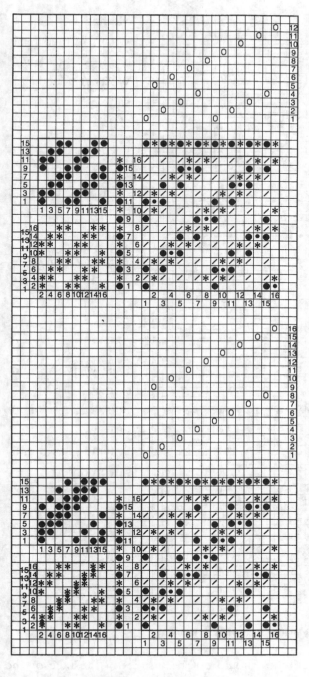

Method I.
Warp: 1 face end — 1 back end.
Weft: 1 face pick — 1 back pick.

Face weave: $\frac{2}{2}$ Z twill.

Back weave: $\frac{2}{2}$ Z twill.

Stitching in 8 end sateen order, count 5.
Face and back weave are extended to 4 repeats to accommodate 8 end sateen stitching.

Face weave: $\frac{3\ 1}{1\ 3}$ Z twill.

Back weave: $\frac{2}{2}$ Z twill.

Stitching in $\frac{1}{7}$ twill order.

Fig. 7.5 Construction of stitching double weave (2) (Method Ⅰ)

Method Ⅰ: <u>warp stitching</u>, <u>back cloth stitched to face cloth</u>.   经接结/下结
Cross-section: face weave = end 1, back weave = end 2, arrows =   上
stitching.
The back ends are lifted above the face picks between floats of lifted face

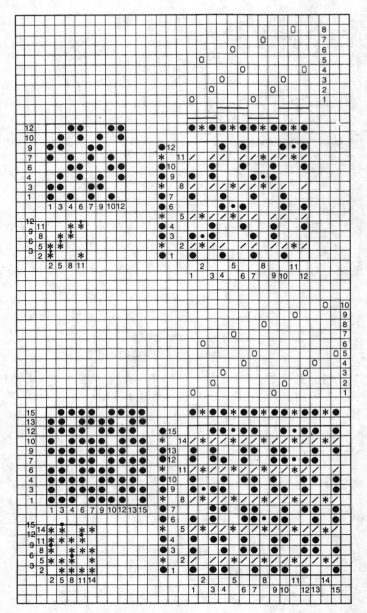

Method I.
Warp: 1 face — 1 back - 1 face end.
Weft: 1 face — 1 back — 1 face pick.
Face weave: $\frac{2}{2}$ broken twill.
Back weave: $\frac{2}{2}$ Z twill.
Stitching: in $\frac{1}{3}$ twill order.

Face weave: 5 end satin.
Back weave: 5 end satin.
Stitching: in 5 end sateen order.

Fig. 7.6 Construction of stitching double weave (3) (method I)

ends on either side. These stitches should not be visible on the face of the fabric.

Development of structure.

F = face weave: $\frac{2}{2}$ Z twill.

B = back weave: $\frac{2}{2}$ Z twill, the stitches are marked with dots on the lines between back picks 2, 4, 6, 8.

These lines represent face picks 1, 3, 5, 7.
The stitching is arranged in twill order.
T = Transferring the weaves.
Warp: 1 face end - 1 back end. Weft: 1 face pick - 1 back pick.
Repeat: 8 ends/8 picks.
Face on ends and picks 1, 3, 5, 7. Back on ends and picks 2, 4, 6, 8.
Warp stitching on ends 8, 2, 4, 6. picks 1, 3, 5, 7.
Lifter: all face ends are lifted over back picks.

7.2.2 Face cloth stitched to back cloth

Method Ⅱ: weft stitching, face cloth stitched to back cloth.   纬接结/上接

Cross-section: face weave = end 1, back weave = end 2, arrows =   下
stitching.

The face ends are lowered below back picks between lowered back ends on either side. The stitches should not be visible on the back of the fabric.

Development of structure.

F = face weave: 2/2 Z twill, the stitches are marked with circles (canceling out the lifter) on the lines between face picks 1, 3, 5, 7.   去掉

These lines represent back picks 2, 4, 6, 8. The stitching is arranged in twill order.

B = back weave: 2/2 Z twill.
T = Transferring the weaves.

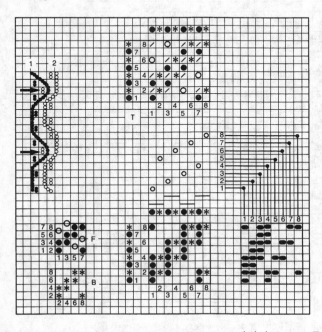

Fig. 7.7 Construction of stitching double weave (4) (Method Ⅱ)

Warp: 1 face end - 1 back end. Weft: 1 face pick - 1 back pick.
Repeat: 8 ends/8 picks.
Face on odd ends and picks. Back on even ends and picks.
Weft stitching on ends 5, 7, 1, 3. picks 2, 4, 6, 8.
Lifters: face ends are lifted over back picks,
circles = no lifts (cancelled lifters).

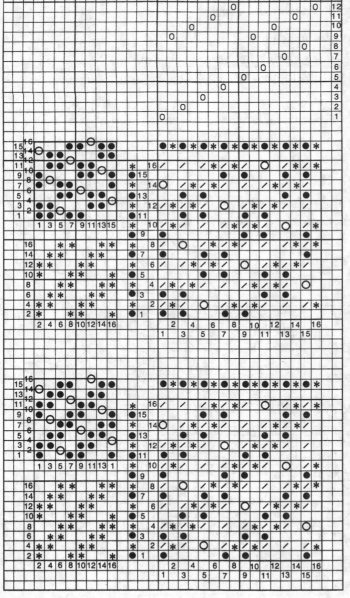

Method II.
Warp: 1 face end - 1 back end.
Weft: 1 face pick - 1 back pick.

Face weave: 8 end twilled hopsack.
Stitching: 8 end sateen order.
Back weave: $\frac{2}{2}$ Z twill.

Face weave: $\frac{2}{2}$ Z twill.
Stitching: 8 end sateen order.
Back weave: $\frac{2}{2}$ Z twill.

Fig. 7.8　Construction of stitching double weave (5) (Method II)

Less coherent stitching has been applied to these examples.

粘接

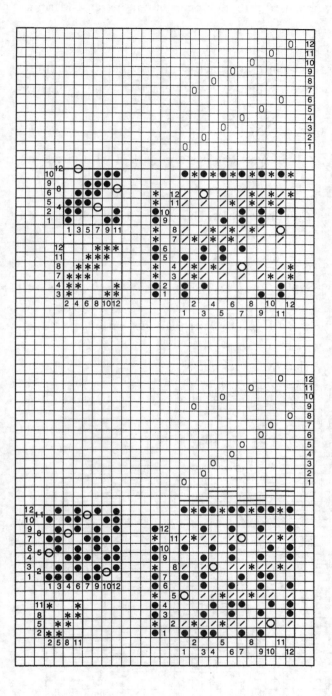

Method Ⅱ.
Warp: 1 face — 1 back end.
Weft: 2 face — 2 back picks.

Face weave: $\frac{3}{3}$ Z twill.
Stitching: only every second end is stitched.

Back weave: $\frac{3}{3}$ Z twill.

Warp: 1 face — 1 back — 1 face end.
Weft: 1 face — 1 back — 1 face pick.

Face weave: 8 end crepe.
Stitching: only every second end is stitched.
Back weave: $\frac{2}{2}$ Z twill.

Fig. 7.9 Construction of stitching double weave (6) (Method Ⅱ)

上下对接

7.2.3 Double stitching

Both series of threads can be used for stitching to obtain increased firmness of structure.

Method Ⅲ: weft and warp stitching.

Face cloth stitched to back cloth and back cloth stitched to face cloth.

Cross-section: face weave = end 1, back weave = end 2, arrows = stitching.

Development of structure.

F = face weave: $\frac{2}{2}$ Z twill weft stitches are marked with circles on the lines between face picks.

B = back weave: $\frac{2}{2}$ Z twill, warp stitches are marked with dots on the lines between back picks.

T = Transferring the weaves.

Face: on ends and picks 1, 3, 5, 7. Weft stitches on ends 5, 7, 1, 3, picks 2, 4, 6, 8.

Back: on ends and picks 2, 4, 6, 8. Warp stitches on ends 8, 2, 4, 6, picks 1, 3, 5, 7.

Lifters: face ends are lifted over back picks,

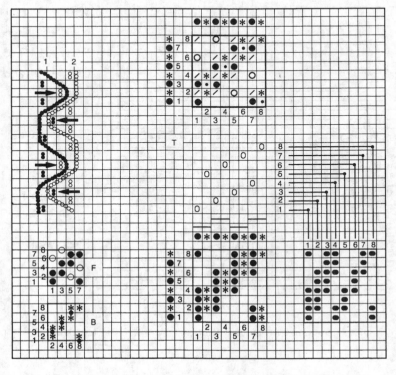

Fig. 7.10　Construction of stitching double weave (7) (Method Ⅲ)

circles indicate cancelled lifters.

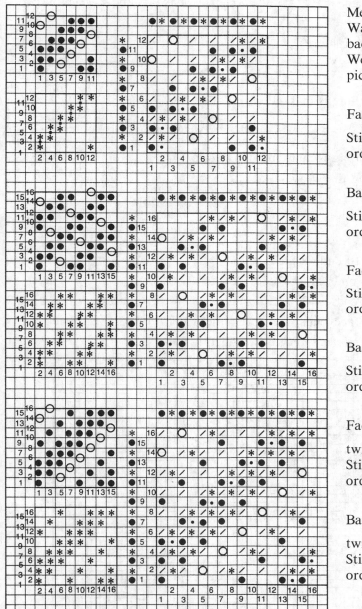

Method Ⅲ.
Warp: 1 face-1 back end.
Weft: 1 face-1 back pick.

Face: $\frac{3}{3}$ Z twill.
Stitching in twill order.

Back: $\frac{2}{2}$ Z twill.
Stitching in twill order.

Face: $\frac{2}{2}$ Z twill.
Stitching in sateen order, count 5.

Back: $\frac{2}{2}$ Z twill.
Stitching in sateen order, count 5.

Face: $\frac{3\ 1}{1\ 3}$ Z twill.
Stitching in twill order.

Back: $\frac{3\ 1}{2\ 2}$ Z twill.
Stitching in twill order.

Fig. 7.11　Construction of stitching double weave (8) (Method Ⅲ)

7.2.4 Center warp stitching (Fig. 7.12, Fig. 7.13)　　接结经

If the colours or counts of the yarns for the face and back cloth differ greatly fine centre threads can be use for stitching.

Method Ⅳ: face and back cloth are stitched with a centre warp.

Due to the different take up of face and back warp a separate beam is required.

Cross-section: face weave = end 3, centre weave = end 4, back weave = end 5.

Development of structure.

F = face weave: $\frac{2}{2}$ Z twill. Lines between face ends represent centre ends 4 and 9. Stitching is marked with dots.

B = back weave: $\frac{2}{2}$ Z twill. Lines between back ends represent centre ends 4 and 9. Stitching is marked with circles.

T = Transferring the weaves.

Warp: face 1 1.1 1. = 4. Weft: 1 face — 1 back pick.

back 1.1 1.1 = 4.

centre. 1.. 1. = 2.

Repeat: 10 ends/8 picks.

Face: on ends 1, 3, 6, 8. Picks 1, 3, 5, 7.

Back: on ends 2, 5, 7, 10. Picks 2, 4, 6, 8. Centre: on ends 4 and 9.

Lifters: face ends over back picks centre ends over back picks, circles indicate cancelled lifters.

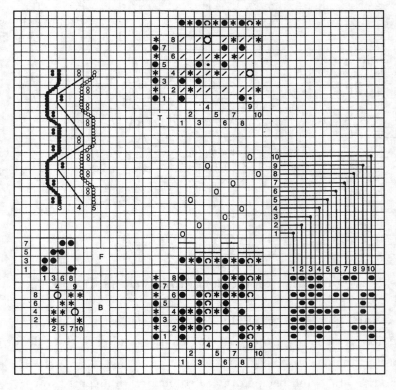

Fig. 7. 12 Construction of stitching double weave (9) (Method Ⅳ)

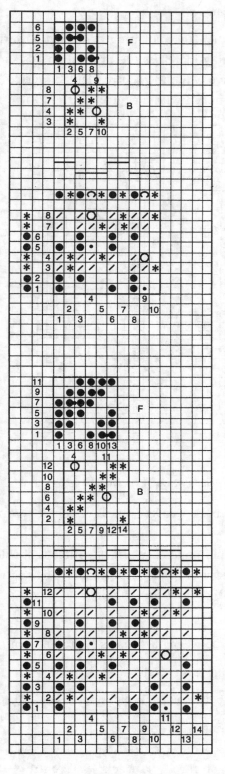

Method IV.

Face weave: $\frac{3}{1}$ Z twill.

Back weave: $\frac{2}{2}$ Z twill.

Warp: face 1 1.1 1. = 4.
　　　back 1.1 1.1 = 4.
　　　centre. 1.. 1. = 2 (4 and 9).
Weft: 2 face picks – 2 back picks.
Repeat: 10 ends/8 picks.

Face weave: $\frac{4}{2}$ Z twill.

Back weave: $\frac{2}{4}$ Z twill.

Warp: face 1 1.1 1 1.1 = 6.
　　　back 1.1 1 1.1 1 = 6.
　　　centre. 1... 1.. = 2 (4 and 11).
Weft: 1 face pick – 1 back pick.
Repeat: 14 ends/12 picks.

Fig. 7. 13 Construction of stitching double-weave (10) (Method IV)

接结纬

7.2.5 Center weft stitching (Fig. 7.14, Fig. 7.15)

This type of stitching is only applied when the mounting of an extra warp beam is not possible. Centre weft stitching increases the production costs as the take up has to be rendered inoperative while the centre pick is inserted.

Method Ⅴ: face and back cloth are stitched with a centre weft.

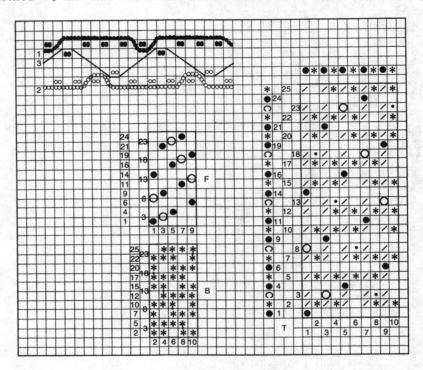

Fig. 7.14 Construction of stitching double weave (11) (Method Ⅴ)

Cross-section: face weave — pick 1, back weave — pick 2, centre weave = pick 3.

Centre picks are stitching over face and under back ends. Stitches should be hidden between adjacent face or back weft floats respectively. The centre threads lie between face and back when not used for stitching.

Development of structure.

F = face weave. $\frac{1}{4}$ sateen. Lines between face picks represent centre picks, mark stitching with circles.

B = back weave. $\frac{4}{1}$ satin. Lines between back picks represent centre picks, mark stitching with dots.

Transferring the weaves.

Warp: 1 face — 1 back end. Weft: 1 face — 1 back — 1 centre — 1 face — 1 back pick.

UNIT Ⅱ　Compound Structure　**133**

Repeat: 10 ends/25 picks.
Face weave: on ends and picks marked • .
Back weave: on ends and picks marked ＊.
Centre on picks 3, 8, 13, 18, 23.
Interlacing: back ends over centre picks (dots),
　　　　　　face ends below centre picks (circles).
Lifters: face ends over back and centre picks, circles are cancelled face lifters.
　As with centre warp stitching, the yarn for centre weft stitching should be finer in count than either face or back and <u>neutral</u> in colour, especially when the colours for the face and back cloths differ greatly. This also helps to <u>conceal</u> the stitches.

中性

掩盖

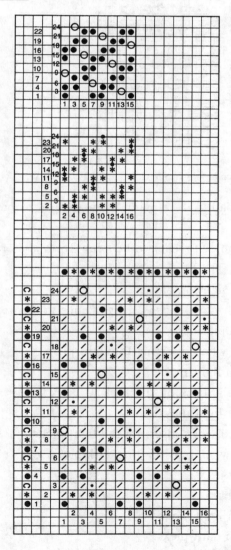

Method Ⅴ.

Face weave: $\frac{2}{2}$ Z twill.

Stitching picks: weft stitching (circles) in sateen order.

Back weave: $\frac{2}{2}$ Z twill.

Stitching picks: warp stitching (dots) in sateen order.

Warp: 1 face end — 1 back end.
Weft: 1 face-1 back — 1 centre pick.
Repeat: 16 ends/24 picks.

Fig. 7.15　Construction of stitching double weave (15) (Method Ⅴ)

7.3 Interchanging double cloth

This is a technique in which two series of warp interlace with two series of weft to produce simultaneously two layers of fabric.

The object usually is to create a cloth in which colours interchange to give solid coloured, mixed coloured effects.

The two layer of fabric exchange with each other to form the character of the design. It is on these points of interchange that the two cloths are bound together. See Fig. 7.16.

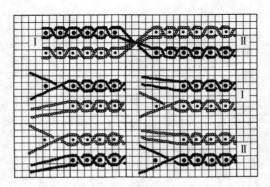

Fig. 7.16 Interchanging double cloth

The end uses of these fabric are manifold, ranging from apparel to home textiles and furnishing.

The two layer of fabric exchange with each other can form different effects. See Fig. 7.17 effects Ⅰ, Ⅱ, Ⅲ, Ⅳ. F—face, B—back.

Ⅰ: Solid dark colour on the face, light on back.
   Face: dark ends interlace with dark ends, all light ends are lowered when inserting the dark picks.
   Back: all dark ends are raised when inserting the light picks, light ends interlace with light picks.

Ⅱ: Solid light colour on the face and dark on the back.
   Face: light ends interlace with light picks, all dark ends are lowered when inserting the light picks.
   Back: all light ends are raised when inserting the dark picks, dark ends interlace with dark picks.

Ⅲ: Mixed colour on the face and back.
   Face: dark ends interlace with light picks.
   Back: light ends interlace with dark picks.

Ⅳ: Mixed colour on the face and back.

Face: light ends interlace with dark picks.
Back: dark ends interlace with light picks.

The steps of development of interchanging double cloth as following:

(1) Design the patterns.

(2) Selected the base weaves which the simple weaves are preferable due to the low number of shafts required.

(3) Colour arrangement: the ratios are usually 1 : 1, 2 : 1, 2 : 2.

(4) Determine the size of the pattern, the repeat must be times of the base repeat.

(5) Draw the weaves in each parts of the pattern.

Different effect needs different weave. See Fig. 7.17.

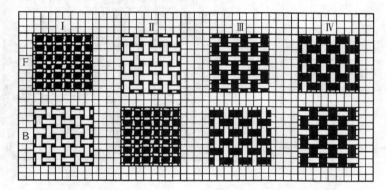

Fig. 7.17　Double weave interchanging effects

Example 1: see Fig. 7.18.

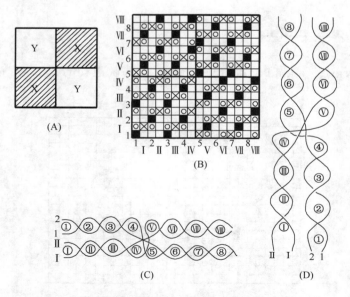

Fig. 7.18　Interchanging double weave

(A) is the pattern. Base weaves are plains.
Arabic 1, 2, 3, 4, ⋯, indicate X colour.
Roman Ⅰ, Ⅱ, Ⅲ, Ⅳ, ⋯, indicate Y colour.
The repeat:

$$R_0 = R_y = 2 \times (4 \times 2) = 16.$$

According to the weave and cross-section and longitudinal-section of the fabric, we can easily find that the weave and arrangement are meet the pattern requirement.

Example 2: see Fig. 7.19. Four effects are achieved.

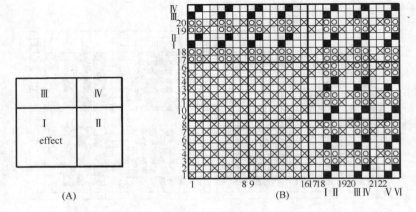

Fig. 7.19 Interchanging double weave

7.4 Tubular cloths

Tubular weave is a kind of double weave which both selvedges are joined. See Fig. 7.20.

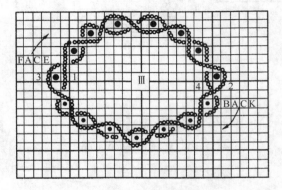

Fig. 7.20 Tubular cloth

Tubular fabrics are used for fire hoses, seamless bags, sacks, tubular shaped filter, covering Cylindrical objects, artificial vessels, etc. Various dimensions can be produced.

无缝袋子/袜子/管状过滤/柱体外罩/人造血压管

In order to get a perfect tubular fabric. Certain rules for the design of the fabric have to be observed.

7.4.1 The main points of designing a tubular fabric

(1) Selection of base weaves. The face weave and back weave should be same. And the structure should be simple. The following weaves can be used as the bases: plain weave, $\frac{2}{2}$ weft ribs, twill, etc. The shift in weft direction should be constant; otherwise, the selvedges are not evenness.

整齐

(2) Arrangements of face and back threads. Arrangements in warp direction should be 1:1, and in weft direction must be 1:1; otherwise, the selvedges can not be joined properly.

连接

(3) Calculation of the total number of ends. To achieve a perfect continuation of the weave from face to back and vice versa, certain rules for the calculation of the total number of ends in fabric have to be observed.

Here, the formula is recommended.

$$M_{0t} = R_0 \times Z \pm S_y,$$

here: M_{0t}—total number of ends;
R_0—base weave repeat;
Z—the number of base weave repeats;
S_y—shift in weft direction.

If the base weave is plain, the M_{0t} should be odd.
If the first pick passing from left to right, S_y is negative.
If the first pick passing right to left, S_y is positive.

(4) Drawing the weave. The cross section of the tubular fabrics is helpful for drawing the weave. See Fig. 7.21.

7.4.2 The main points need to be paid attention for weaving plan

(1) Drafting: It is better using straight drafts or grouped drafts. If grouped drafts are used, face ends should be dawn in front shafts.

(2) Denting: To counteract the tendency of the weft to contract more at the sides than in the centre of the cloth, the first and last three dents in the reed should each have approximately one third less threads than are drawn into the other dents.

(3) Cords: In order to achieve an evenness selvedges, the selvedge cords are used which are thick threads indented to prevent an increase of warp density near the selvedges. The cords are lifted in such a manner that they

特线

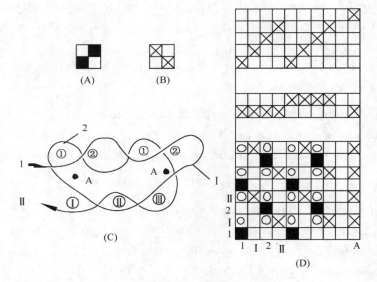

(A)—face weave
(B)—back weave
(C)—cross section of the tubular fabric
(D)—weaving plan

Fig. 7.21 Construction of tubular weave

are not woven in the fabric. These cords will removed when the fabric is taken out of the loom.

7.5 Double width cloths

Sometimes, we need the width of fabric wider than the width of customary loom. We can use the principles of double weave to construct a double width fabric. Double width fabric is a kind of double weave which one selvedge is joined. See Fig. 7.22.

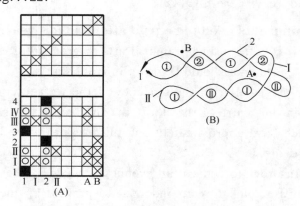

Fig. 7.22 Construction of double width weave

Double width fabrics are widely for industrial purpose, such as <u>paper-making blanket</u>. 造纸毛毯

7.5.1 The main points of designing a double width cloths

(1) Base weave selection. Base weave should be simple, such as plain, twills, and regular satin/sateen. Nevertheless, the base weave must meet the requirement of the end product.

(2) Arrangement of face and back threads. Arrangement in warp way of face and back can be 1∶1, 2∶2. The best way is to choose 1∶1. Arrangement in weft way of face and back must be 2∶2.

(3) The first insertion is limitation in direction. See Fig. 7.22.

Here: The base weave is 1/1 plain;
　　　Arrangement in warp is 1∶1;
　　　Arrangement in weft is 2∶2;
　　　A—selvedge cord;
　　　B—stitching yarn to achieve an evenness selvedge which removed after woven.

7.5.2 The main points for weaving plan (Fig. 7.23—Fig. 7.26)

(1) The materials, count, weave of the face and back should be same. Therefore, one beam is adaptable to construct the cloths.

(2) Either one shuttle or multi-shuttle are suitable.

(3) Using grouped draft or straight draft. Grouped draft can easy get evenness tension of warp ends.

(4) <u>Selvedge cord</u>. <u>Stitching yarn</u> are used to achieve an evenness selvedges. 特线/缝线

7.6 <u>Multi-layer weaves</u> 多层组织

7.6.1 <u>Treble Cloths</u> (1) 三层

A treble cloth is composed of three series of warp and three series of weft threads. One series of each kind is forming the <u>face, centre and back fabric</u>. 表/中/里

Cross-sections Ⅰ and Ⅱ: face weave = end 1, centre weave = end 2, back weave = end 3.

Examples Ⅰ and Ⅱ.

Warping and picking order = 1 face - 1 centre - 1 back.

Structures of face, centre and back = 2/2 Z twill.

Method Ⅰ: centre cloth is stitched to face cloth. <u>Stitches</u> are marked with 接结点
dots, back cloth is stitched to centre cloth. Stitches are marked with dots.

Method Ⅱ: face cloth is stitched to centre cloth. Stitches are marked with circles, centre cloth is stitched to back cloth. Stitches are marked with circles.

经组织点　　Circles cancel out lifters.

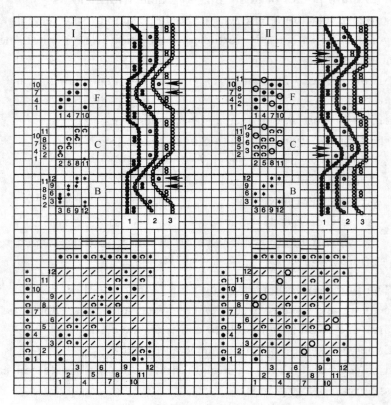

Fig. 7.23 Construction of treble cloths

7.6.2 Treble cloths (2)

Cross-sections Ⅲ and Ⅳ: face weave = end 1, centre weave = end 2, back weave = end 3.

Examples Ⅲ and Ⅳ.

Warping and picking order = 1 face - 1 centre - 1 back.

Structures of face, centre and back = 2/2 Z twill.

Method Ⅲ: face cloth is stitched to centre cloth. Stitches marked with circles.

back cloth is stitched to centre cloth. Stitches marked with dots.

Method Ⅳ: centre cloth is stitched to face cloth. Stitches marked with dots.

centre cloth is stitched to back cloth. Stitches marked with circles.

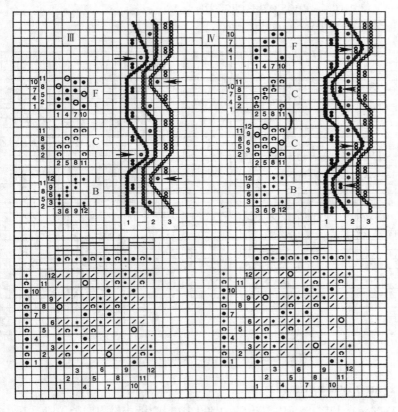

Fig. 7.24 Construction of treble cloths

7.6.3 Four-ply cloths(1)   四层织物

A four-ply fabric is composed of four series of warp and four series of weft threads. One series of each kind is forming the face, centre Ⅰ, centre Ⅱ and back fabric.   表/中Ⅰ,/中Ⅱ/里

In this example all fabrics interlace in 2/2 Z twill.

Cross-section: face = end 1, centre Ⅰ = end 2, centre Ⅱ = end 3, back = end 4.

Warping and picking order: face = ends and picks 1, 5, 9, 13;
centre Ⅰ = ends and picks 2, 6, 10, 14;
centre Ⅱ = ends and picks 3, 7, 11, 15;
back = ends and picks 4, 8, 12, 16.

Stitching: only every second end of each fabric layer is stitched.

Face to centre Ⅰ (circles), centre Ⅰ to centre Ⅱ (circles), centre Ⅱ to back (circles).

Repeat: 16 ends/16 picks.

Lifters: face ends over centre Ⅰ, Ⅱ and back picks,
centre Ⅰ ends over centre Ⅱ and back picks,

centre II ends over back picks.
Circles indicate cancelled lifters.

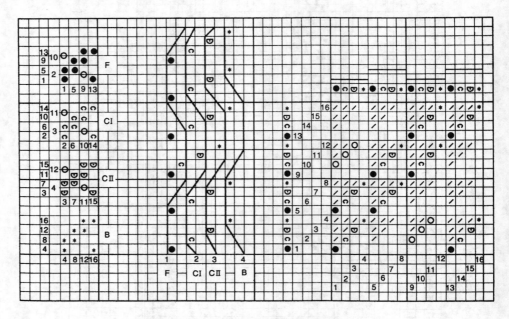

Fig. 7.25 Construction of four-ply cloths

多层组织

带子

7.6.4 Four-ply Cloths (2)

These types of fabrics are mainly used for <u>belts</u> and <u>straps</u>. Layers that are very tightly interlaced with each other allow only for a limited amount of picks to be beaten up.

In this example all fabric layers interlace plain weave.

Cross-section: F = face weave, C I and C II = centre weaves, B = back weave.

All ends and picks are numbered.

Warp: face = ends 1, 5, centre I = ends 2, 6, centre II = ends 3, 7, back — ends 4, 8.

Weft: face 1.. 1 = 2/picks 1 and 8;
　　　centre I 1.1. = 2/picks 2 and 7;
　　　centre II 1 1.. = 2/picks 3 and 6;
　　　back 2... = 2/picks 4 and 5.

Stitching:

Face to centre I (circle), centre I to centre II (circle), centre II to back (circle).

Centre I to face (dot), centre II to centre I (dot), back to centre II (dot).

Repeat: 8 ends/8 picks.

Lifters: face ends over centre Ⅰ, Ⅱ and back picks,
 centre Ⅰ ends over centre Ⅱ and back picks,
 centre Ⅱ ends over back picks.
Circles indicate cancelled lifters.

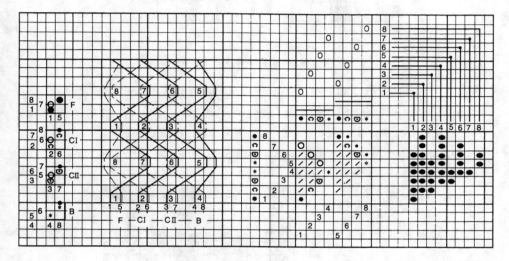

Fig. 7.26 Construction of four-ply cloths

7.6.5 Multi-layer weaves

The industrial textiles get more and more important as the application fields of textiles expanded. The three-dimensional woven textiles are widely used. Multi-layer woven fabrics are important of three-dimensional woven textiles.

Method description

Take a multi-layer of woven fabrics which the each layer are the same as an example to describe the design method of multi-layer woven fabrics.

(1) Determine the number of layers of the required fabric as n.

(2) The weaves named sectional-weaves are gained as A1 and A2, and the repeat of the weaves A1 and A2 are abstained by the formula:

$$R_0 = R_y = n,$$

here: R_0 ——warp repeat of sectional-weaves,
 R_y ——weft repeat of sectional-weaves.

The warp floats of each space of the sectional-weaves A1 and A2 are as follows respectively:

A1: $n, n-1, n-2, \cdots, 2, 1$;
A2: $n-1, n-2, \cdots, 2, 1, 0$.

Take a 4-layer sectional-weaves A1 and A2 as an example shown in Fig. 7.27.

(3) Select the base weaves for each layer of the multi-layer woven fabric,

such as plain, twill.

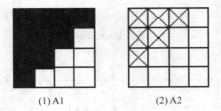

(1) A1   (2) A2

Fig. 7. 27 Sectional weaves of 4-layer weave

(4) Replacement: The new weave of the multi-layer woven fabric is obtained by replacement based on the selected base weaves, that each warp float is replaced by A1, and each weft float is replaced by A2.

A non-stitched 4-layer plain weave and a non-stitched 4-layer twill weave are shown in Fig. 7.28.

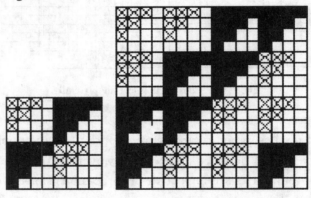

(1) 4-layer plain weave   (2) 4-layer twill weave

Fig. 7. 28 4-layer plain weave & 4-layer twill weave without stitching

Example

A nine-player plain woven fabric is taken for example as shown in Fig. 7.29. It is found that is easier to draw the weave.

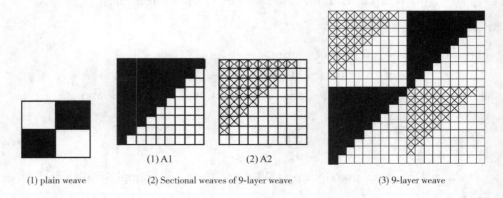

(1) plain weave   (2) Sectional weaves of 9-layer weave   (3) 9-layer weave

(1) A1   (2) A2

Fig. 7. 29 9-layer weave based on plain weave

Steps for drawing the multi-layer weaves as following:
(1) Draw the base weave;
(2) Draw the sectional weaves;
(3) Draw the multi-weave by replacement.

Stitched multi-fabrics

多层接结织物

Warp stitching can be get by adding warp floats, and weft stitching by removing warp floats, that is adding weft floats. An intensive warp stitching multi-layer weave is shown in Fig. 7.30, that is to increase the thickness of the fabric. Another patterned warp stitching multi-layer weave is shown in Fig. 7.31, and its cross-section diagram shown in Fig. 7.32. From Fig. 7.31 and Fig. 7.32, it can be seen that the warp thread in the second layer interlaced with weft thread in the first layer, the fourth layer interlaced with the third layer, the third layer interlaced with the second layer, and the fifth layer interlaced with the fourth layer. The size of the cellule is directly depended by n. The various pattern stitching forms many different performs of textile reinforced <u>composite materials</u>.

复合材料

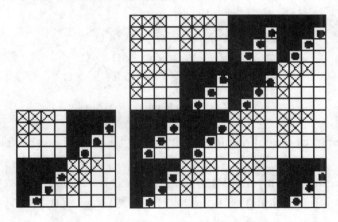

(1) 4-layer stitched plain weave (2) 4-layer stitched twill weave

Fig. 7.30 Weave diagram stitched from lower to upper

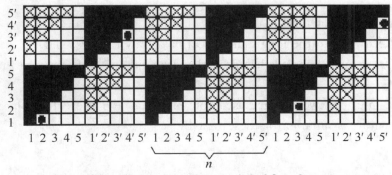

Fig. 7.31 Weave diagram with pattern stitched from lower to upper

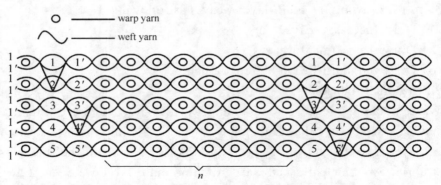

Fig. 7.32 Cross-section diagram with pattern stitched from lower to upper

7.6.6 Angle-interlock fabric

Angle-interlock fabric is a special multi-layer fabric which has one system of warp, and multi-system of weft. The warp interlaces with each weft. See Fig. 7.33. This fabric structure not only can make different thickness of fabric, but also make the fabric easy bending.

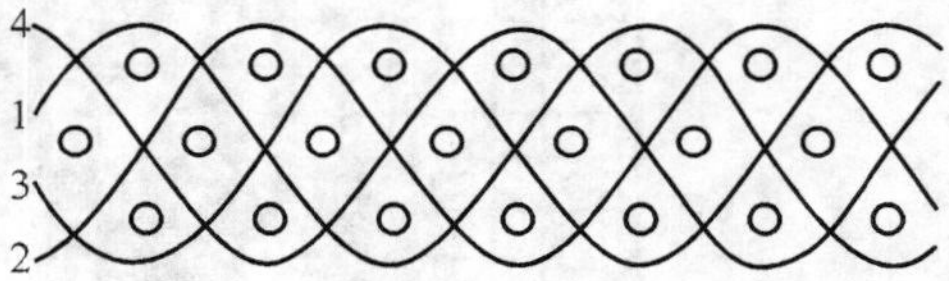

Fig. 7.33 Angle-interlock Weaves

Development of angle-interlock weaves can be carried out by the procedure below.

(1) Determine the layers. Fig. 7.34 shows 2-ply, 3-ply, 4-ply cloths.

(2) Draw a longitudinal section. See Fig. 7.34 (A), (B), (C).

(3) Calculate the repeat R_0, R_y and shift S_0.

$R_0 = P$ (the number of ply) $+ 1$,

$R_y = R_0 \times P = P(P+1)$,

$S_0 = P$.

The maximum of the float $f_m = 2P - 1$.

(4) According to the longitudinal section, draw the first end.

(5) Draw the other ends based on the first end and shift S_0. See Fig. 7.34 (D), (E), (F).

Angle-interlock weaves are used for industrial and protective placations. Straight draft is suitable for these weaves.

UNIT Ⅱ Compound Structure   **147**

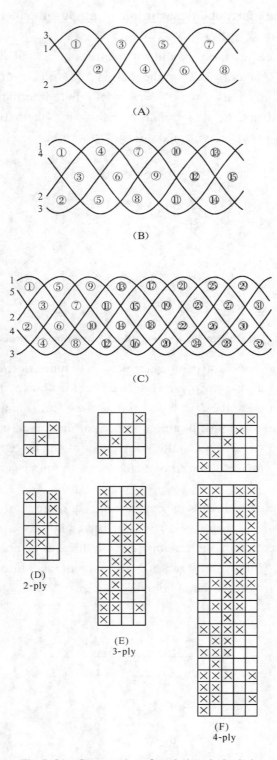

Fig. 7.34 Construction of angle-interlock cloths

角联锁织物
分工表示法

7.6.7 Fractional formula description of angle-interlock woven fabric method description

The above-described method is valid for the less layer number angle-interlock fabric, such as 4 layers. But in the design of large layer number angle-interlock fabric, such as 9 or 10 layers of angle-interlock weaves, it is difficulty to construct the weave plan by using the schematic diagram as in Fig. 7.35, and it is also easy to make mistakes in practice.

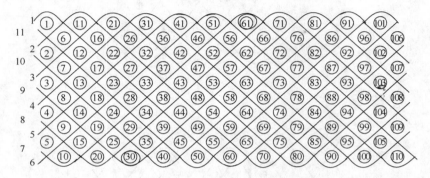

Fig. 7.35 The schematic diagram of 10-layer angle-interlock woven fabric

It is found that each warp interlacement is symmetric in angle-interlock weave. So as long as one warp yarn interlacement is known, the interweaving of other warp yarns can be drawn according to the shift number of each warp yarn. Because the relationships among other parameters of construction have been given from Eqs. (1)—(4), the weave plan of the large number layers of angle-interlock fabric could be constructed when the interlacing rule has been characterized.

We number the weft threads in sequence of requirement of weaving process as shown in Fig. 7.35, which is based on the tight beat-up of weft yarns. From the investigation of interlacing rules of the odd layers and even layers respectively, the following fractional formula of the interlacing between warp and weft yarns could be obtained.

对偶数层 For even layers:

$$\frac{a}{b} = \frac{\left(\frac{P}{2} - i\right), \cdots, 1, 0, j, \cdots, \frac{P}{2}}{j, \cdots, \frac{P}{2}, \left(\frac{P}{2} - i\right), \cdots, 1, 0},$$

$$i = 0, 1, 2, \cdots, \frac{P}{2},$$

$$j = 0, 1, 2, \cdots, \frac{P}{2},$$

i, j: repeat one time and not repeat only when $j = 0, \frac{P}{2}$.

For odd layers:

$$\frac{a}{b} = \frac{\left(\frac{P+1}{2}-i\right), \cdots, 1, 0, j, \cdots, \frac{P-1}{2}}{j, \cdots, \frac{P-1}{2}, \left(\frac{P+1}{2}-i\right), \cdots, 1, 0},$$

$$i = 0, 1, 2\cdots, \frac{P+1}{2},$$

$$j = 0, 1, 2, \cdots, \frac{P-1}{2},$$

i, j: repeat one time and not repeat only when i, $j = 0$.

Here: a: number of intersection points of warp yarn;
b: number of intersection points of weft yarn;
P: layers of angle-interlock woven fabric;
i, j: natural number.

Method used

If a 10-layer angle-interlock weave construction is needed drawing, we can get the first warp yarn interlacement with the above-mentioned fractional formula as follows.

The first warp yarn interlacement is:

$$\frac{5\ 5\ 4\ 4\ 3\ 3\ 2\ 2\ 1\ 1\ 0\ 0\ 0\ 1\ 1\ 2\ 2\ 3\ 3\ 4\ 4\ 5}{0\ 1\ 1\ 2\ 2\ 3\ 3\ 4\ 4\ 5\ 5\ 5\ 4\ 4\ 3\ 3\ 2\ 2\ 1\ 1\ 0\ 0}.$$

It can be seen from the above fractional formula that there are 55 warp intersection points and 55 weft intersection points respectively. The weft repeat is 110 which meets Eq. (2): $R_y = P \cdot (P+1) = 10 \times 11 = 110$. And the longest warp floats and weft floats are both 19 as Eq. (4): $F_m = 2P - 1 = 20 - 1 = 19$.

When the interlacing rule of the first warp yarn is known, the interlacing rules of other warp yarns can be easily obtained in term of the shift number of warp yarn.

Another 9-layer angle-interlock weave, for instance, is drawn as following by the fractional formula method.

The first warp yarn interlacement is:

$$\frac{5\ 4\ 4\ 3\ 3\ 2\ 2\ 1\ 1\ 0\ 0\ 0\ 1\ 1\ 2\ 2\ 3\ 3\ 4\ 4}{0\ 1\ 1\ 2\ 2\ 3\ 3\ 4\ 4\ 5\ 4\ 4\ 3\ 3\ 2\ 2\ 1\ 1\ 0\ 0},$$

$R_0 = P + 1 = 9 + 1 = 10,$
$R_y = P(P+1) = 9 \times 10 = 90,$
$S_0 = P = 9,$
$F_m = 2P - 1 = 2 \times 9 - 1 = 17.$

The weaving plan is shown in Fig. 7.36 and schematic diagram of 9-layer angle-interlock woven fabric is shown in Fig. 7.37. They are both conformed to the formula.

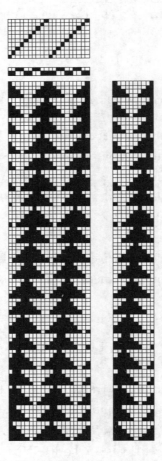

Fig. 7.36　Weaving plan of 9-layer angle-interlock

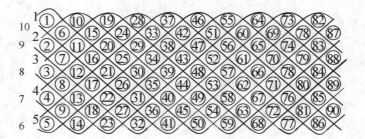

Fig. 7.37　Schematic diagram of 9-layer angle-interlock woven fabric

HOMEWORKS

1. Construct the following double weaves.
   (1) Face weave $\frac{2}{2}\nearrow$ twill, back weave $\frac{2}{2}\nearrow$ twill, arrangement 1 : 1.
   (2) Face weave $\frac{2}{2}\nearrow$ twill, back weave $\frac{1}{1}\nearrow$ plain, arrangement of face and back in both directions are 2 : 1.
2. Construct the following stitching weaves.
   (1) A coating cloth, face weave $\frac{2}{2}\nearrow$, back weave $\frac{1}{1}$ plain, arrangement $m : n = 2 : 1$, stitching in plain order. (use Method I)
   (2) A coating cloth, face weave $\frac{2}{2}\nearrow$, back weave $\frac{2}{2}\nearrow$, arrangement $m : n$ in warp way 1 : 1, in weft way 2 : 2, stitching in $\frac{1}{3}$ twill order (use Method I)
   (3) Face weave $\frac{2}{2}$ hopsack, back weave $\frac{2}{2}\nearrow$, arrangement $m : n = 1 : 1$, stitching in $\frac{1}{3}$ twill order (use Method II) and draw the draft and lifting plan.
3. Construct a interchange double weave, face and back weave $\frac{1}{1}$ plain, arrangement, 1B 1L the pattern as Fig. 7.38.

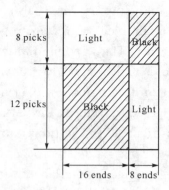

Fig. 7.38

4. Construct a tubular weave, the base weave is $\frac{2}{2}$ weft rib, and draw the cross section.

Chapter Eight
Pile Fabrics
绒织物

Pile fabrics are characterized by the brush-like surface formed by tufts of warp or weft cut threads. Cutting the threads can be done either on the loom or on the machines of the fabric finishing department. Such fabrics should be distinguished from the others which receive a pile after passing through a raising machine or after electrical flocking.

Pile fabrics can be divided into the following two groups.

1. Weft pile fabrics, containing a weft cut pile produced by cutting the weft threads at the fabric finishing department by specially constructed knives.

2. Warp pile fabrics, having a pile produced on the loom from the threads of an extra warp called a pile warp.

8.1 Weft pile

Weft pile fabrics. For producing the weft pile fabrics, two systems of weft threads and one system of warp threads are necessary. If the pile weft differs in count or colour from the ground weft, the loom should be equipped with a multi-shuttle mechanism, and, in some cases, also with a dobby.

These fabrics usually contain a much greater proportion of weft threads as compared to the warp threads. The weft pile fabrics are known as velveteen and corduroy. Velveteen is uniformly covered with a short dense pile. Corduroy has longitudinal cords, or ribs, on the surface formed by the tufts of different length. In figured velveteens the tufts of pile are arranged in such a manner as to form figures. In the process of cloth formation on the single-shuttle loom, the weft plays a double role.

First, it forms the fabric ground by interlacing with warp threads in a plain or twill weave, and second, while interlacing with the warp it forms long floats of weft threads, which will be cut to form a pile in finishing.

The length of long floats determines the height of the pile.

The main representative products are corduroy and velveteen.

8.1.1 Corduroy

Corduroys posses <u>soft handle</u>, <u>clear stripes</u>, <u>lofty</u> and <u>full pile</u>, good strength, and <u>aesthetic properties</u>, so they are widely used for apparel fabrics. They are favourited by man, woman, old and children.

灯芯绒
手感柔软/绒
条清晰/膨松/
丰满面满/美观

8.1.1.1 The principles of construction of corduroy fabric

See Fig. 8.1. The weave contains two <u>pile picks</u> a, b, and two <u>ground picks</u> 1, 2. The ground picks 1, 2 interlace with warp forming the <u>ground weave</u>. Arrangement of ground pick and pile pick is 1∶2. The pile pick floats over five warp threads which are convenient for <u>cutting process</u>. The pile picks interlace with the warp threads 5, 6 for binding the piles. The intersections are called <u>pile roots</u>.

毛纬/地纬
地组织

割绒

绒根

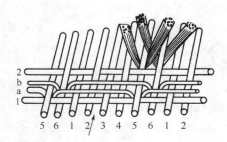

Fig. 8.1 Sketch of corduroy

After weaving, the pile weft was cut by a special constructed knives between the warp threads 2 and 3. Then, brushing it, the piles will be upright, forming a full pile, bulky band corduroy.

The principles of the cutting are described as following. See Fig. 8.2. The <u>circular knife</u> are placed on a mandrill A indicated by the arrow.

圆刀

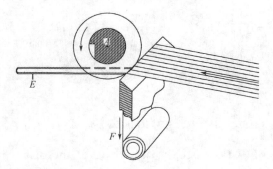

Fig. 8.2 Sketch of corduroy cutting

As the knives revolve, fabric advances toward in the direction indicated by an arrow F.

A <u>guide wire</u> E is inserted in the fabric under the long weft floats. The guide wires have these functions: ① guiding the weft floats forming a loop-

导针

like: "race" to the knives, and tautening them as they are cut; and ② keeping the knives in the centre of each "race". As the uncut fabric approaches the knives, guide wires are conveyed along by it, and consequently require to be pushed forward again intermittently. This is accomplished by a series of spirally arranged rotary cams.

8.1.1.2 The classification of corduroy

According to the width of cord, the corduroys are classified as following: ① fine (needle) corduroy, more than 11 cords/25mm; ② mid-wale corduroy, 9—11 cords/25mm; ③ wide-wale corduroy, 6—8 cords /25mm; ④ spacious-wale cording, less than 6 cords/25mm.

According to the ways of finishing process, the corduroys are classified as following: ① printed corduroy; ② dyed corduroy.

According to the materials used, the corduroys are classified as following: ① cotton corduroy; ② blended corduroy.

8.1.1.3 The weaves and setts

① Density: warp cover factor suit 50%—60%, weft cover factor suit 140%—180%; and the ratio between warp cover factor and weft cover factor suit 1/3. ② Ground weave: the ground weave's function is to hold the piles, the widely used weaves are as following, plain, 2/1 twill, 2/2 twill, 2/2 weft rib, 2/2 warp rib. Different ground weaves used can achieve different handle, fastness of the piles. ③ Selection of pile weaves.

Three aspects should be considered for selecting pile weave.

(a) The binding types.

See Fig. 8.3 (A) and (B). There are two types of binding, V-type and W-type.

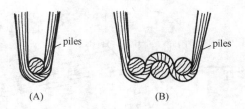

Fig. 8.3 Binding types

The advantage of V-type is easy tighten, and the disadvantage is that the piles are easy being taken out. The opposite of the W-types is.

(b) The height of the piles.

The height of the piles are decided by the length of the pile floats.

$$h(\text{mm}) = \frac{C}{2 \times \frac{P_0}{10}} \times 10 = \frac{50C}{P_0}.$$

UNIT Ⅱ　Compound Structure　155

Here: *C*—the number of the pile floats;
　　　P_0—the warp density.
（c）Distribution of pile roots　　　　　　　　　　　　　　　绒根分布
The distribution of the pile roots decide the appearance of the fabric. Some examples are shown below（See Fig. 8.4）.

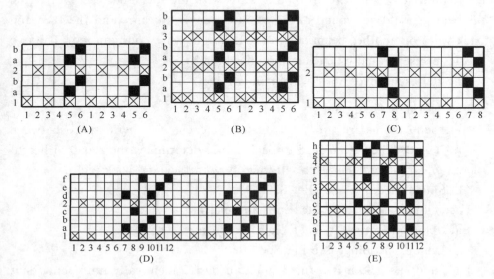

Fig. 8. 4　Corduroy weaves

8.1.2　Velveteen　　　　　　　　　　　　　　　　　　　　纬平绒

If the pile roots are dispersed, the fabric would be covered by an evenness piles. Those fabrics are called velveteens. Velveteens are always <u>raising finished</u> instead of <u>cutting</u>.　　　　　　　　　　起绒整理
　　　　　　　　　　　　　　　　　　　　　　　　　　　　　　割绒
Here is an example. See Fig. 8.5.

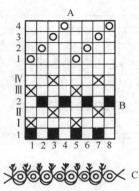

Fig. 8. 5　Velveteen weave

经起毛织物 ## 8.2 Warp pile

绒经 Warp pile fabric contains three systems of warp and two systems of weft. The face and back fabrics are stitched on the loom by means of pile warp, i.e. by interlacing the threads of this warp with the face and back weft. The
经纱用量 pile warp should be wound on a separate weaver's beam, and the face and back warp on another beam due to the different warp consumption. The two fabrics are separated by cutting the pile threads.

经平绒 The representative fabric warp pile is velvet.

8.2.1 Method of weaving

地布 Two ground cloths are produced on top of each other with space between them. The distance from face to back fabric determines the required height of the pile. Pile ends interlace alternately with face and back picks.

单梭口 （1）Single shedding. See Fig. 8.6 (A).

Face cloth is formed by face ends and picks.

Back cloth is formed by back ends and picks.

双梭口 （2）Double shedding. See Fig. 8.6 (B).

Face and back cloth is formed simultaneously, identically interlacing ends
双眼综丝 are drawn into two mail healds.

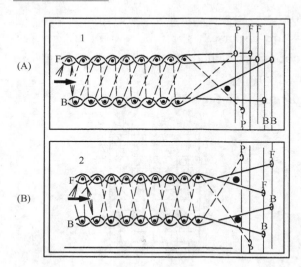

Fig. 8.6 Warp pile weave

Face and back picks are inserted in pairs.

8.2.2 Cutting (arrows)

The two fabrics are separated by cutting the pile threads in the middle during the weaving process. Cutting takes place when the race is put back and the pile ends are taut. Each cloth is to be wound on separately.

Diagrams: F = face, B = back, P = pile.

8.2.3 Development of pile weave

(1) Selection of the ground weave. For velvet, 1/1 plain is preferable. Sometimes, 2/2 weft rib, and 2/1 varied rib are selected. The density of the piles, distribution of the piles need be considered during selection.

(2) Determine the ways of binding type. V-type is preferable than W-type.

(3) Determine the ratio of the ground and pile warp.

(4) Draw a section diagram to help the weave construction.

Two examples are shown below.

Examples for single shed constructions (see Fig. 8.7)

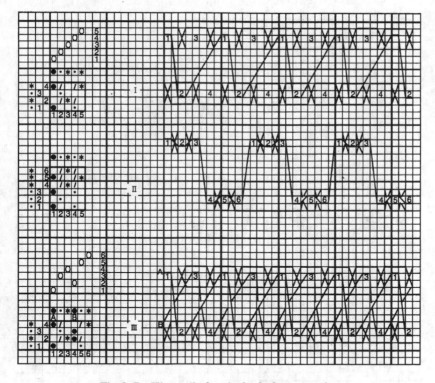

Fig. 8.7 Warp pile for single shed construction

Ⅰ V-type structure, <u>one pile system.</u> 一个绒经系统

Warp: 1 pile end — 1 face — 1 back — 1 face — 1 back end for ground.

Weft: 1 face — 1 back pick.

Ground: plain weave for face and back cloth.

Pile: interlaces alternately with face and back picks.

两个绒经系统

II Fast or W pile structure, one pile system.
   Warp: 1 pile end — 1 face — 1 back — 1 face — 1 back end for ground.
   Weft: 3 face picks – 3 back picks.
   Ground: rib for face and back cloth.
   Pile: interlaces alternately with 3 face and 3 back picks.

III V-type structure, two pile system.
   Warp: 1 pile end A — 1 face — 1 back end.
         1 pile end B — 1 face — 1 back end.
   Weft: 1 face — 1 back pick.
   Ground: plain weave for face and back cloth.
   Pile: both pile ends interlace alternately with face and back picks.
   Cross-sections: numbers represent picks.

Examples for double shed constructions (Fig. 8.8).

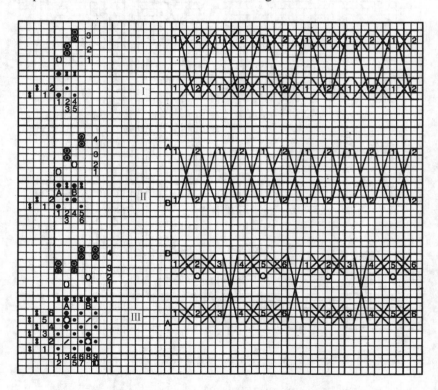

Fig. 8.8 Warp pile for double shed construction

Face and back ends are drawn into two mail healds.
Face and back picks are inserted in pairs.
The weft for examples I, II and III is 1 face — 1 back pick.

I V-type structure, one pile system.

Warp: 1 pile end — 1 face — 1 back — 1 face — 1 back end.
Pile: interlaces alternately with face and back picks.

Ⅱ V-type structure, two pile system.
Warp: 1 pile end A — 1 face — 1 back end
      1 pile end B — 1 face — 1 back end.
Pile: both ends interlace alternately with face and back picks.

Ⅲ Fast pile structure, two pile system.
Warp: 1 face — 1 back end — 1 pile end A — 1 face — 1 back end
      1 face — 1 back end — 1 pile end B — 1 face — 1 back end.
Pile: Lifters (/) over the back picks.

Lowering under the face picks (circles) is indicated in the diagram and cross-section. On these points the ends are moved into a centre position of the shed.

Cross-sections: numbers represent picks. For examples Ⅱ and Ⅲ only the interlacing of the pile ends is drawn out, the plain weave ground being as shown in example Ⅰ.

Warp pile fabrics possess full piles. Soft hand, good elastic recovery, good crease resistance, thermal insulation and delicate colour. They are widely used for lady and children cloth, drapery, seats covers, package cloth for some valuables.

绒毛丰满/手感柔软/弹性好/抗折皱/保暖/色泽柔和

8.3 Terry pile

毛圈织物

The fabric of such weaves are characterized by formation of loops which are raised above the surface of the fabric either on one or both sides.

The fabrics possess good wettability, hygroscopicity, and soft handle. They are used for towelling as well as for bath mats, bed covers and dress.

可湿性/吸湿性

The principles of loops formation are described as following.

8.3.1 Special beating up — terry pile motion

毛圈打纬

During weaving process, several picks of weft are inserted at a short distance from the "fell" of the cloth to produce a short gap; after the last pick is inserted, they are all pushed forward together to their final place in the fabric. Those motions of beating are also called "short beating" and "fast beating".

短打纬/长打纬

8.3.2 Special warp threads

Two systems of warp are essential, ground beam and pile beam. The ground warp have a higher tension which interlaces with weft to form the ground structure; and the pile warp has a loose tension which interlaces with weft to

地经纱织轴/毛经纱织轴

form the loops. The loops are firmly held in the ground structure.

8.3.3 Special weaves and coordination

In order to successfully complete the last fast beating, structurally, three factor are needed to consider.

(1) Resistance of the beating-up should be minimum.

(2) The pile warp is firmly held by ground weave.

(3) After last beating-up, the last weft can not move reversely.

Here, take a three picks terry fabric for example to analyze these three factor. See Fig. 8.9.

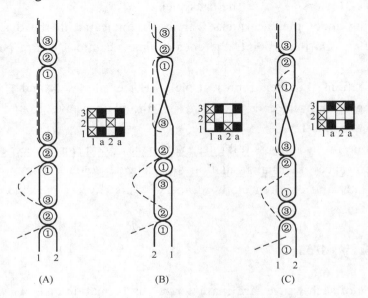

Fig. 8.9 Analysis of three picks terry pile weave

First, resistance of the beating-up.

See Fig. 8.9 (A), (B), (C). At (A), the ground ends interlace with weft two times; and at (B), (C) one time. So weave (A) have a bigger resistance than (B) and (C). (B) and (C) are preferable.

Second, holding force.

(B) and (C) have the same gripping force to the pile warp. At (C), wefts 1 and 2, hold the pile warp; at (B) wefts 2, 3 hold the pile warp. At (B), the loops can not be evenness. The (C) is the best one.

Third, reverse motion of the last pick.

At diagram (A) the weft 3 and next repeat weft 1 are in the same fell. So after the reed go back, the pick 3 will be moved reversely.

At diagram (B), pick 3 can be moved, but it is not serious. The loop would change as the pick remove. At diagram (C), even the pick 3 can be moved, but the gripping force come from wefts 1, 2, so it will not affect the pile loop, and the loop are kept constantly.

According to the analysis above, (C) is the best weave for 3-picks terry fabric.

Examples:
Terry pile (1) (Fig. 8. 10)

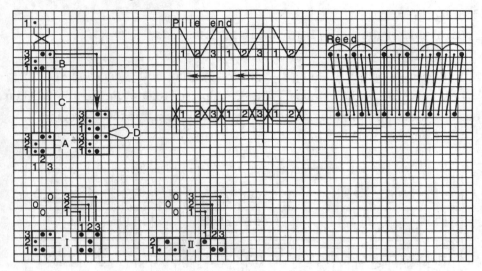

Fig. 8. 10   Terry pile

Single sided pile: Face = pile. Back = ground weave.   单面毛圈

Method of weaving three pick structures.   三纬结构

Warp: 1 ground—1 pile—1 ground end. Weft: solid.   地 经／毛 经／
                                                                                                                        地经
The weave repeat A consists of 3 picks. The ends interlace in rib. Picks 1 and 2 (loose picks) are positioned at a certain distance from the fell of the cloth (fall back), leaving a gap C of approximately 15mm according to the required height of the loop.   毛圈高度

After insertion of pick No. 3 (fast pick) group B is then pushed on to the fell, closing the gap.

For this beating up procedure the pile warp is released. Because the pile   放松
warp is structurally interlaced with the picks it slides along the very taut   拉紧的地经
ground warp, pushing a predetermined length of pile warp forward to form loop D on the face of the fabric.

The loops are held firmly in the cloth by the ground weave. During full   长打纬
beat up the ground ends cross for pick No. 1 of the following group to avoid
a sliding back of previously inserted picks.   返回

Cross-section: the beat up is marked between picks 3 and 1 (arrows).

Method of denting

The very taut ground ends move due to the shrinkage of the cloth to the inside of each dent.

To prevent migrating of pile and ground ends the following order of denting is used.

Left side of the cloth: 1 pile end - 2 ground ends.

Centre: one dent with 1 pile - 2 ground - 1 pile end; this reverses the denting order.

Right side of the cloth: 2 ground ends - 1 pile end.

Ⅰ Three pick structure with draft and lifting plan.

Ⅱ <u>Pile-less boarder heading</u>. A portion of the cloth (end and start of a towel) is woven without loops. The beat up and release mechanism is then out of action. A <u>cramming device</u> increases the picks for these headings.

Terry pile (2) (Fig. 8.11)

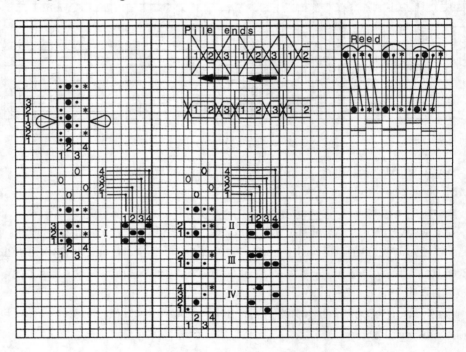

Fig. 8.11 Terry pile

<u>Double sided pile</u>: pile on face and back.

Method of weaving three pick structures.

Warp: 1 ground end — 1 pile end face — 1 ground end — 1 pile end back.

Weft: solid.

The weave repeat consists of 3 picks. The ends interlace in rib.

Pile end 2 forms loops on the face. Pile end 4 forms loops on the back.

Cross-section: the beat up is marked between picks 3 and 1 (arrows).

Denting

UNIT Ⅱ　Compound Structure　163

Left side of the cloth: 1 pile end — 1 ground end.
Centre: one dent with 1 pile — 1 ground — 1 pile end.
Right side of the cloth: 1 ground end — 1 pile end.
According to the warp sett the pile ends can be separated from the ground ends by using a double reed.

Ⅰ: Three pick structure with draft and lifting plan.

Ⅱ，Ⅲ，Ⅳ: Pile-less boarder headings. Various structures can be used according to weft material. For example plain weave or 2/2 interlacing for finer yarns or 1/3 interlacing for coarser weft.

Colouring: face pile and back pile threads can be different in colour.

Terry pile（3）（Fig. 8.12）

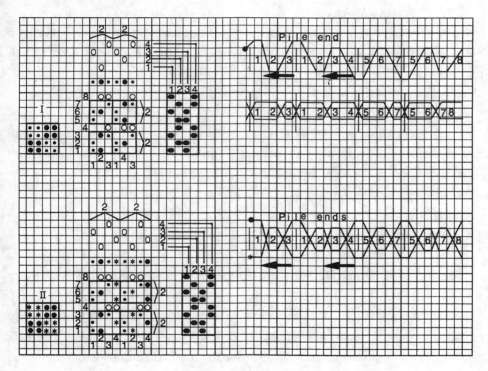

Fig. 8.12　Terry pile

Three pick structures.

Ⅰ Warp: 1 ground — 1 pile — 1 ground end. Weft: solid.
　　　　Ground and pile weave in rib.
　　　　Pile ends change according to the <u>motif</u> from face to back and 图案
　　　　vice versa. The transfer of the pile ends from face to back and
　　　　vice versa requires extra picks（4 and 8）to secure the loops
　　　　properly in the weave.

Cross-section: <u>alternate areas</u> of pile and ground weave form checks on 交换处

both sides of the fabric. The beat up is marked after every third or fourth pick (arrows).

Design/Colours: apart from the exchange of pile variation in the design, different colours within the pile and also the ground can be used. To enlarge designs each section of the draft and weft interlacing has to be repeated to achieve the required size of the figure.

Ⅱ Warp: 1 ground — 1 pile col. 1 — 1 ground — 1 pile col. 2. Weft: solid.

Ground and pile weave in rib.

Pile ends change according to the motif from face to back and vice versa. Extra picks 4 and 8 secure the loops properly in the ground weave.

Cross-section: alternate areas of different coloured pile form checks on both sides of the fabric. The beat up is marked after every third or fourth pick (arrows).

Terry pile(**4**)(Fig. 8.13)

Four pick structure.

With this method a better pattern change is obtained and more intricate designs can be woven. A higher number of picks are necessary (approximately 33 percent more than the three pick structure).

Warp: 1 ground — 1 pile col. 1 — 1 ground — 1 pile col. 2.

Weft: solid.

Ground weave: 2/2 rib.

Motif: steep diagonal.

Weave A: pile colour 1. Weave B: pile colour 2 (the lifts of weave A are reversed).

For the construction of a four pick structure it is necessary that the pile ends form a double pick between every beat up (×). In this construction six pile ends of colour 1 and 2 interlace identically and are numbered accordingly.

Cross-section: alternate areas of different coloured pile form the same pattern on both sides of the fabric.

During the transfer of loops from face to back and vice versa the picks interlace with pile ends in plain weave; the following loop is floating only over 2 picks instead of 3. The beat up is marked after every fourth pick (arrow). Highly picked fabrics can be fully beaten up on the third and fourth pick.

UNIT Ⅱ Compound Structure   **165**

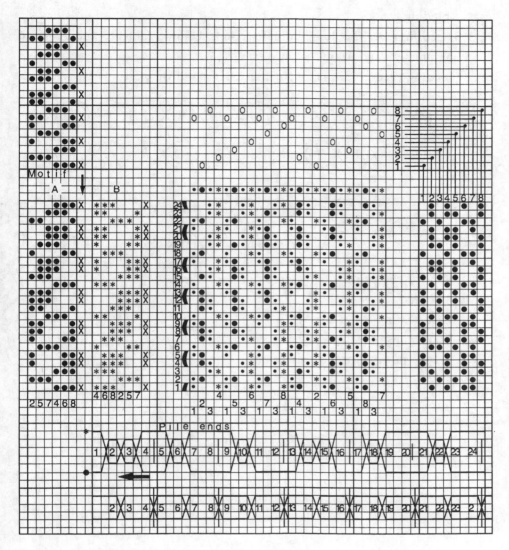

Fig. 8.13 Terry pile

HOMEWORKS

1. List the binding types of corduroy, and compare the difference of each type.
2. Construct the following corduroy weaves.
   (1) Base weave $\frac{2}{1}\nearrow$, ground pick: pile pick $= 1:2$, $R_0 = 6$, V-type binding.
   (2) Base weave $\frac{1}{1}$ plain, ground pick: pile pick $= 1:2$; $R_0 = 8$, W-type

binding.

3. According to the longitudinal-sections below, indicate the required parameters and draw the weaves.

(1) binding types; (2) arrangement; (3) ground weaves; (4) weaves.

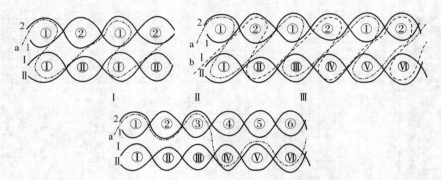

Ⅰ   Ⅱ   Ⅲ

4. Describe the principles of terry pile formation.

Main points:

(1) Fast beating and short beating;

(2) Loose tension warp, higher tension warp;

(3) Weave possess three requirements.

Chapter Nine
Gauze and Leno Weaves
纱罗组织

9.1 The concept of gauze

基本概念

These weaves differ from the others in that the warp threads of one system do not run parallel or at right angles with the weft threads, but are twisted round them. 扭绞

The fabrics of suchweaves are used for dress, curtains, decorative fabrics, and mosquito nets. 蚊帐布

The characteristic feature of these fabrics is the spaces between the threads, which produce an open net-like structure. The weft threads are firmly held by the crossing warp threads, ensuring a uniform texture. These firm open fabrics, are also used for technical purposes, such as for the selvedges on the shuttle-less looms and for producing the pile yarn, so-called chenille weft. 纱孔 / 雪尼尔线

9.2 The principles of gauze weave formation

纱的形成原理

9.2.1 Some new terms (see Fig.9.1, Fig.9.2)

Doup thread—also called crossing thread, during the weaving process, the doup threads are raised in each shed, and it passing each time under the standard threads forms a zig-zag line. 绞经 / 地经

Standard or regular thread—during weaving, it remains comp- aratively straight.

Standard shaft—the shaft is as same as conventional shaft. 后综

Doup shaft—also called lifting heald. 半综、绞综

Lifting healds L1 and L2 consist of two individual flat steel strips. Close to their centre they are connected to each other to form a resting point (:) which is necessary for the lifting of the doup heald D. Doup head shanks slide between the steel strips. The bottom end of each shank is fitted with a carrying rod. These two rods are connected to each other at their end by stoppers and constitute the doup heald frame. 焊点 / 腿 / 托体

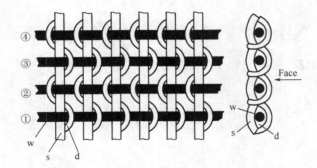

Fig. 9.1 Gauze weave, plan and section

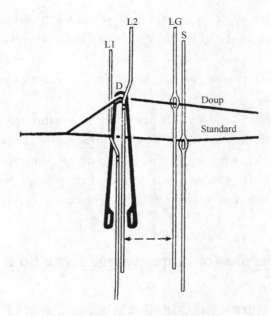

Fig. 9.2 Mechanism of gauze weave

9.2.2 Method of weaving

地经/绞经   Two kinds of warp threads are necessary: <u>standard ends</u> and <u>doup ends</u>.

地综   The doup end is drawn into the heald of the <u>ground shaft</u> LG and into the eyelet of the doup heald D.

绞综   The standard end is drawn into the heald of standard shaft S and between the space of the <u>lifting healds</u>.

The doup heald frame with the lifting heald shaft is fitted at the front.

Gauze ground shaft LG and standard shaft S are placed with a gap of approximately 8–12cm behind the doup heald frame. This space is necessary to allow for a reasonable formation of the crossed shed.

The mechanism of shedding system of gauze or leno can form three shed as following: see Fig.9.3 ①,②,③.

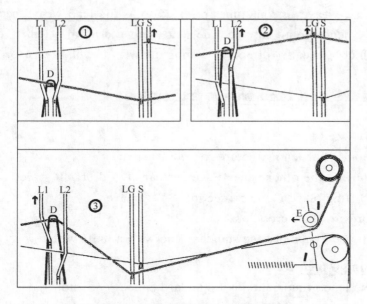

Fig. 9.3 Mechanism of gauze weave formation

9.2.2.1 <u>Plain shed</u>   普通梭口

Only the standard end is raised by the standard shaft S.

Shafts LG, L1 and L2 with doup heald frame remain lowered.

This plain shed forms part of the ground weave. It is used for leno weaves.

9.2.2.2 <u>Open shed</u>   开放梭口

The doup end is raised by shafts LG and L2 with doup heald D on the right side of the standard end. Shafts L1 and S are lowered.

This open shed forms the other part of the ground weave. Standard ends have free passage between the upper part of L1 and the raised doup heald D. See Fig. 9.1, gauze weave weft ① can be woven by this shed.

9.2.2.3 <u>Crossed shed</u>   绞转梭口

The doup end raised by shaft L1 with doup heald D on the left hand side of the standard end. Shafts L2, LG and S are lowered. See Fig. 9.1, gauze weave weft ② can be woven by this shed.

In the crossed shed position (heavy lift) the <u>easer bar</u> E is pulled towards   松经棒
the harness because the tension of doup ends must <u>be slackened</u> so that the   放松
threads do not have to carry <u>excessive strain</u>.   过度拉伸

There are a number of easer motion devices which can be used, but they depend on the construction of the loom and more particularly on the dobby. In order to obtain a clear shed the <u>doup ground shaft</u> is kept about 12mm   地综
lower than the standard shaft.

More than one beam is generally necessary, due to the different take up of the threads. How many beams depends on the design effect required.

纱/纱罗
纱罗织机

The gauze weave is sometimes referred to as the leno weave because it is made on a leno loom, but the true leno weave is a variation of the gauze weave which is usually formed by combining with the plain weave.

纱罗组织示例

9.3 Gauze and leno weave examples

Example 1（see Fig. 9. 4）:

Development of leno structures on point paper.

Diagrams on the other page: E = easer bar. D = draft. R = reed. S = structure and lifting plan. C = cross-section. W = weave.

Structures for half cross leno.

The leno end encircles the standard end with a half twist.

单纬纱罗

9.3.1 Single pick leno

Warp: the leno group contains 1 standard end-1 leno end.

Draft: the leno end is on the right side of the standard end and then crosses through the doup heald to the left side of the standard end.

Reed: one leno group has to be drawn into the same space between the dents.

Weave: pick 1 = open shed. Pick 2 = crossed shed, easer bar moved forward. The leno end crosses after each pick below the standard end from one side to the other but interlaces over every pick. The standard ends remain below each pick.

单纬对绞纱罗

9.3.2 Single pick counter leno

Warp: leno group one = 1 standard end-1 leno end.
　　　　leno group two = 1 leno end-1 standard end.

Draft: in the first leno group the leno end is on the right side of the standard end and then crosses through the doup heald to the left of the standard end.

In the second leno group the leno end is on the left side of the standard end and then crosses through the doup heald to the right of the standard end.

This resembles a reversed or point draft arrangement.

Weave: pick 1 = open shed. Pick 2 = crossed shed, easer bar moved forward. The two neighboring leno groups bind simultaneously, but in opposite directions. Leno ends cross after each pick below the standard ends from one side to the other but interlace over each pick. Standard ends remain below each pick.

Reed: each group is drawn into the same dent, but may be separated by an empty dent.

9.3.3 Multi pick counter leno

多纬对绞纱罗

Warp and draft as Ⅱ.

  Weave: pick 1 = open shed. Picks 2 and 4 = plain shed.
         Pick 3 = crossed shed, easer bar moved forward.

Apart from the single pick weaves where the leno end changes after each pick to the other side of the standard end, there are multi-pick leno weaves in which two or more picks are inserted before the leno end changes to the other side of the standard end.

In this example the standard ends are lowered on all uneven picks and raised over all even picks.

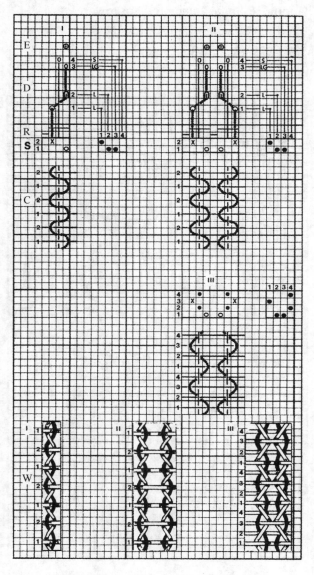

Fig. 9.4 Gauze and leno weave, plan

Leno fabrics（1）

普通绞综

With eyelet doup healds（see Fig. 9.5）

Combined single and counter leno with twill stripe.

Draft：

斜纹条带

Shafts 9—12 = <u>twill stripe</u>.

Shafts 1，2，5，7 = group one, leno ends guided over bar E 1.

Shafts 3，4，6，8 = group two（shafts 6，8 in reversed order）. Leno ends guided over bar E 2.

Weave：

Group one：single pick leno, picks 1，3，5，7 are crossed sheds, easer 1 in action. Standard ends remain lowered.

Group two：multi pick counter leno, picks 1，2，7，8 are crossed sheds, easer 2 in action. 4 picks are inserted before the leno ends change to the other side of the lowered standard ends.

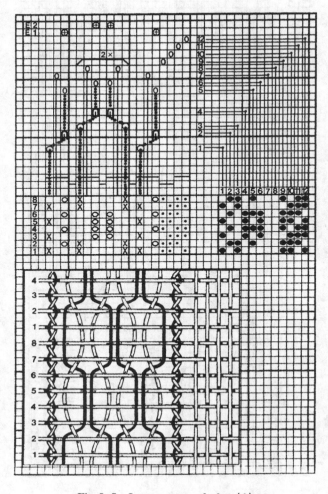

Fig. 9.5 Leno weave and plan（1）

Leno fabrics(2)

With eyelet doup healds(see Fig. 9.6) 普通绞综

Counter leno with two standard ends in each group.

Draft:

Shafts 8, 9 = plain weave stripe.

Shafts 10—13 = 3/1 twill stripe. 斜纹条带

Shafts 1, 2, 3, 4, 5 = group one, all leno ends are guided over E.

Shafts 1, 2, 3, 6, 7 = group two (shafts 2, 3 in reversed order).

Weave:

Group one/two: counter leno, picks 1—6 are crossed sheds, easer E in action. 6 picks are inserted before the leno ends change to the other side of the plain weave and rib interlacing standard ends.

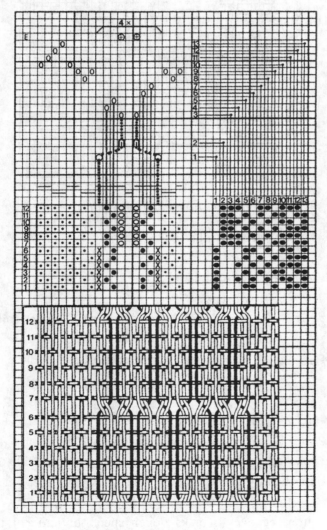

Fig. 9.6 Leno weave and plan (2)

Leno fabrics(3)

With eyelet doup healds (see Fig. 9.7, Fig. 9.8)

Counter leno with two standard ends in each group.

Draft:

Shafts 1, 2, 3, 4, 5—group one. All leno ends are guided over E.

Shafts 1, 2, 3, 4, 5—group two (shafts 2, 3 in reversed order).

Weave:

Picks 5—8 are crossed sheds, easer E in action. 4 picks are inserted before leno ends change to the other side of the plain weave interlacing standard ends.

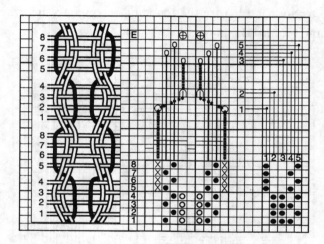

Fig. 9.7 Leno weave and plan (3)

Particulars for the design on the opposite page.

Draft:

Shafts 11, 12 = plain weave stripe.

Shafts 13—16 = 3/1 twill stripe.

Shafts 1, 2, 5, 7, 8 = group one (in reversed order). Leno ends guided over E 1.

Shafts 3, 4, 6, 9, 10 = group two (in reversed order). Leno ends guided over E 2.

Weave:

Picks 5—8 and 21—24 are crossed sheds, easer 1 in action.

Picks 9—12 and 17—20 are crossed sheds, easer 2 in action.

4 picks are inserted before leno ends change to the other side of the standard ends. Standard ends float alternately over or under 4 picks in reversed order.

Leno ends interlace over 4 picks (open shed) or under 4 picks (plain shed).

Standard ends interlace plain weave and twill.

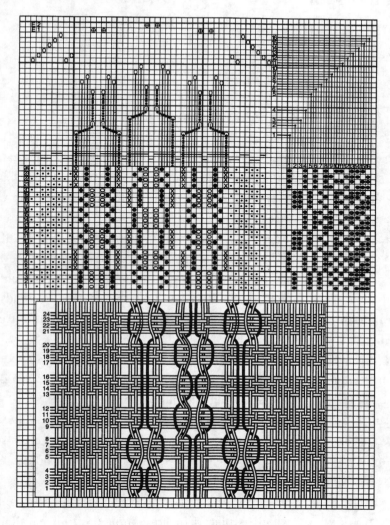

Fig. 9.8　Leno weave and plan (4)

HOMEWORKS

1. Understand the following terms:
   standard thread, doup thread,
   standard shaft, doup shaft,
   ground shaft, leno, open shed, plain shed,
   crossed shed, gauze, counter leno.
2. Understand the weaving process of gauze.

Chapter Ten
Jacquard Fabrics
纹织物

10.1 Elements of jacquard shedding

Large-pattern fabrics can be produced on looms equipped with a particular type of shedding motion-Jacquard machines. These are used for designs with several hundreds of warp threads interlacing in different manner and with the same number of weft threads in the repeat. In some designs the repeat of weave reaches several thousands. In Jacquard machines not only small groups of warp threads are lifted, but single threads as well. It is possible because instead of healed shafts, the harness cords are used, each of which carries one individual warp thread. The lifting and lowering of single threads in different orders make it possible to produce the figures of desirable form and size.

There is a close relation between the Jacquard design and the type of Jacquard machine to be used. For better understanding of the Jacquard designs, some elements of Jacquard shedding should be considered.

The single-cylinder Jacquard machine is shown in Fig. l0.1. The threads are lifted by hooks 1, which are arranged in rows. In the cross-section, where the vertical plane cuts the machine at the short row of hooks, 8 hooks can be seen. There is a greater number of hooks (50 in some cases) in the long row. The hook is made of wire. The top bent part of hook can be engaged by the knife 2, which is placed in horizontal position parallel to the long row of hooks. The number of knives equals the number of the long rows of hooks. The knives are mounted in the knife frame 3, which reciprocates vertically once every pick. The bottom part of hook is bent in the form of loop, and the neck cord 4 is attached to this part. The bottom part is placed over the rod 5. The rods are arranged in the grate 6.

Each hook is connected with horizontal needle 7, which can deviate the hook to the left. Eight needles are placed in the same vertical plane, forming a short row of needles. The right-hand ends of the needles protrude at the side of the machine, passing through the holes of needle board 8, which holds the ends of the needles in the proper position against the holes of cylinder 9.

The left-hand end of the needles is positioned in the spring box 10. The spring 11 is intended to return the needle to initial position after the needle has been deviated by the card 12.

The harness cords are attached to each neck cord 4. They are drawn through the holes of the comber-board which is placed in the horizontal position above the warp parallel to the fabric fell. The length of the comber-board is a little bit greater than the width of the fabric. The comber-board is intended for uniform arrangement of the harness cords.

花板
综线
目板

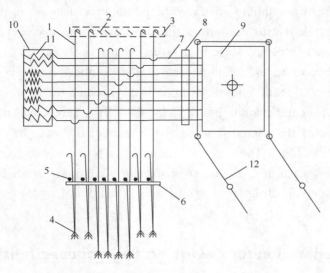

1. hooks 2. knife 3. knife frame 4. neck cord
5. rod 6. grate 7. horizontal needle 8. needle board
9. cylinder 10. spring box 11. spring 12. card

Fig. 10.1 Schematic diagram of Jacquard machine

带钩竖针/提刀/提刀架/龙头绳/托棒/托格/横针/针板/花筒/弹簧盒/弹簧/花板/铅锤

To the bottom end of harness cord, a heald, or mail, is attached, to which a weight or lingoes is suspended. By means of this weight the heald, warp thread, cord and hook are returned to their initial positions after they have been lifted.

At machine operation, the cylinder moves to the left, bringing the card to the needles. And if there is no hole in the card against the needle, the card presses the needle, moves the needle and deflects the hook to the left. The hook will not be engaged by the lifting knife. The warp thread remains in the lower position, forming the bottom part of the shed.

If there is a hole in the card opposite the needle, it is not deflected. The hook remains over the knife and at the next moment, the knife grasps it while going up, and lifts it. As a result, the harness cord and the warp thread are lifted. The warp thread forms the top part of the shed.

Thus, if there is a hole in the card, the warp thread is lifted and overlaps the pick. The number of holes in the card determines the number of warp threads to be lifted in the particular shed, when this card operates. And the number of cards in the chain equals the number of weft threads in the repeat. It should be taken into consideration while constructing the Jacquard design and the card-cutting tables.

The warp repeat of the fabric to be produced depends on the type of Jacquard machine. The number of differently interlacing threads in the fabric can be equal to, but should not exceed, the number of hooks in the machine. It should be taken into account while calculating the necessary number of hooks for producing the fabric of a given warp repeat.

The number of holes per unit area of the comber-board depends on the warp density. The total number of holes is the product of the number of holes in the short row and the number of rows. Usually, there are as many holes in the short row of the comber-board, as the number of hooks in the short row of Jacquard machine. The width of the comber-board equals the warp width in reed. The length of the comber-board can be chosen to satisfy the conditions: the number of holes per unit area should not exceed the standard value.

10.2 Preparation for designing the jacquard fabrics

The starting point of the preparation is to calculate the number of threads in warp and weft repeats of the weave. In this case, the size of the pattern and the warp and weft densities in finished fabric should be given. The warp repeat of the weave can be found by multiplying the warp density by the width of the pattern, and the weft repeat can be found in the same way.

The warp repeat of the Jacquard weave determines the figuring capacity of the Jacquard machine. The total number of hooks should be equal to or more than the warp repeat. When the type of machine has been chosen, it is necessary to find from the machine specification the number of hooks in the short row, which will be used for further calculations.

Jacquard weave is constructed on a sheet of special design paper, which is then used as an instruction for cutting the cards, and it can be called then "a card cutting plan". The design paper is divided by thick lines into square blocks, each of which is subdivided into horizontal and vertical spaces. Each horizontal space corresponds to a weft thread, and each vertical space to a warp thread. Each small square of the block corresponds to one overlap. For convenience, in the designing and card cutting, each block is divided by ver-

tical lines into as many spaces as there are hooks in the <u>short row</u> of the Jac-   横行
quard machine.

A design paper is characterized by the <u>count</u>, which is a ratio of vertical   规格
spaces to horizontal ones in each block. In order to retain correct proportions
and shapes of figure designs, the ratio of vertical spaces to horizontal ones
should be equal to the ratio of warp density to the weft density in the finished
fabric.

<u>Count of design paper</u> can be calculated by the formula,   意匠纸规格

$$a/b = P_0/P_y,$$

here a = number of vertical spaces in the block, which is usually assumed
equal to the number of hooks in short row of Jacquard machine;
b = number of horizontal spaces in the block;
P_0 = warp density of the finished fabric;
P_y = weft density of the fabric.

Next step is to find the number of blocks, which in horizontal direction is
calculated by the formula,

$$M = R_0/a,$$

here R_0 is the warp repeat of the Jacquard weave.

The number of blocks in vertical direction is calculated in the similar way,

$$N = R_y/b,$$

here R_y is the weft repeat of the weave.

The total number of block is $M \times N$.

Now, it is possible to find the size of the necessary sheet of design paper,
taking into account that the width of block equals the length of it, in other
words the sides of a block are of the same value, for instance.

The width of the design paper is calculated by the formula,

$$L_w = l \times M,$$

here M is the number of blocks in horizontal direction.

The length of the design paper can be found by a similar formula,

$$L_e = l \times N,$$

here N is the number of blocks in vertical direction.

The size of the design paper is $L_w \times L_e$.

纹织物设计
步骤

10.3 Steps in construction of jacquard design

10.3.1 Studio techniques/conventional system (Fig. 10.2)

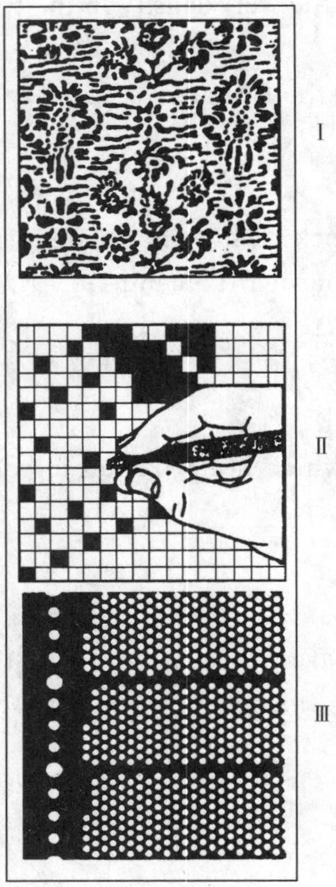

图案设计

Ⅰ. Designer: creating new ranges, pattern sketches.

意匠设计

Ⅱ. Drafter: transferring sketch on to point paper complete with structure.

轧花设计

Ⅲ. Cutter: semi-mechanical cutting methods, reading in weave structure from point paper.

Fig. 10.2 Construction of jacquard design by conventional way

描图纸

The first step is to draw the picture or to copy it from the art book or any other source. A copy can be made by using the tracing paper. The figure of any shape can be chosen, and as an example of figure the flower has been chosen, which is shown at (A) in Fig. 10.3.

The picture should be divided by vertical lines into a certain number of spaces equal to the number of blocks in horizontal direction of the design paper, which have been calculated before constructing the design. Twenty four vertical spaces are marked by digits in the same direction as the blocks at (B) in Fig. 10.3. The tracing paper should be divided by horizontal lines in 14 spaces, because there are 14 blocks in vertical direction on the design paper in Fig. 10.3 (B).

UNIT Ⅱ Compound Structure 181

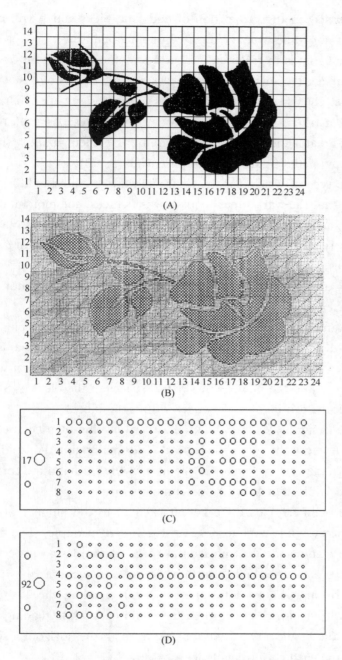

Fig. 10.3 Construction of Jacquard design

Thus, the tracing paper is divided into 336 squares, equal to the number of blocks of design paper, the count of which is 8 : 8. The count of the design paper has been calculated before constructing, taking into account the densities of threads of finished fabric and the number of hooks in the short row of Jacquard machine, which is equal to 8 in this example. The count of the design paper based on the proportion of ends and picks which will ensure that the figure is not to be woven <u>elongated or squashed</u>.

The second step is to transfer the outline of the figure from tracing paper to design paper. It can be done easily, since the number of squares of the tracing paper equals the number of blocks of the design paper. The outline of the figure is marked at (B) in Fig. 10.3 by grey squares. As a rule, the size of the block is greater than that of the square on the tracing paper, therefore an enlarged copy of the figure will be constructed at (B).

Now, the type of weave for the <u>figure</u> and the <u>ground</u> should be chosen. It is advisable that the warp repeat equals or is the least common multiple of the number of small squares in the block in horizontal direction, and that of weft, in vertical direction. The warp overlaps of the figure are marked by black squares, and that of ground-by red dots. However, other methods of constructing are known, where the warp overlaps are either not painted or the whole area of figure is painted.

In this example at (B) in Fig. 10.3 (B) plain weave is chosen for the figure, and twill 1/7 for the ground. The repeat of twill 1/7 is equal to the number of vertical spaces in the block. The least common multiple of warp repeat of plain weave and the number of vertical spaces in the block is 8, and warp repeat of weave can be placed 4 times within the block.

Now, the design is ready to be used as <u>card cutting plan</u>, because it shows the lifting of every thread in the repeat. The small <u>painted</u> or <u>marked square</u> on the design paper <u>corresponds</u> to the hole in the card and the corresponding warp thread is lifted at Jacquard machine operation.

The total number of cards to be cut is equal to the number of horizontal spaces of the design paper, to $8 \times 14 = 112$ in our example. The first card is made while reading the first horizontal space of the design paper. The card cutter reads from left to right, cutting the holes in accordance with sequence of painted or marked by dots squares. While reading the 17th horizontal space of the first block, the <u>card cutter</u> cuts hole 1 while reading the second block also hole 1, and so on, and looking through the horizontal space of the 13th block the card cutter cuts holes 1, 4, 5, 7 (Fig. 10.3 (C)).

While reading the horizontal space 92 (Fig. 10.3 (D)) the card cutter cuts the holes 4, 7, 8, then 1, 5, 6, 8 in the second short row of the card, then 2, 4, 6, 8 in the third and in the fourth then, 1, 4, 5, 7, 8 in the fifth, the hole 2-in the sixth, and so on, until the card is ready.

10.3.2 Electronic data processing system

In Jacquard weaving(Fig. 10.4), computer technology is now successfully employed.

    Designer: creating new ranges. Pattern sketches or designs can be either created on the <u>colour monitor</u> or images from any media can be <u>scanned</u> into the system. Colourways viewed. Designed printed out.   电子直接设计/扫描

    Programmer: <u>structures added</u>, <u>exchanged</u>, <u>repeated</u>, <u>re-scaled</u>, <u>contours adjusted</u>. The information is transferred into the electronically controlled cutting machine or directly record on <u>magnetic tape cartridge</u>.   组织添加/交换/循环计算/比例缩放/边缘修正/磁盘

With the latest electronically controlled Jacquard machine the hooks are replaced by an <u>electromechanical system</u>. It is so efficient that the weaving process can be speeded up to ranges which until now were considered unrealistic. The lifting system of the ends in theory remains the same. The cards however can be replace with a magnetic tape.   机电系统

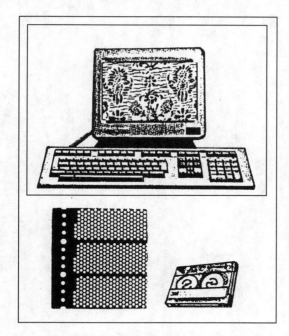

Fig. 10.4 Construction of jacquard design by electronic way

HOMEWORKS

1. Get to know each part of the Jacquard machine at Fig. 10.1.
2. Get to know how the design paper is selected.

UNIT III

Fabric Design
织物设计

Chapter Eleven
Fabric Geometry
织物的几何结构

11.1 Fabric geometry

Studying the relations of fabric parameters, the fabric can be considered as an interlacing of two systems of flexible circular cylinders. The cylinders intersect at right angles. In such a geometric model of fabric the relations between the diameters, spacing and bending are of great importance. Knowing these purely geometric relations, useful conclusions can be given concerning the closest possible spacing, the maximum fabric cover, the yarn crimps and the thickness of fabric.

These data are necessary for the designer of fabrics. Studying the geometry of the fabric, not only the maximum number of threads per cm can be estimated, but also the ratio of densities of warp and weft threads. From fabric geometry it is possible to explain the relations between warp and weft crimps, between relative covers. There is a great influence of parameters of the fabric structure on its properties. The crimp has a marked influence on the elongation and strength, the fabric thickness, the mass per square metre, and so on. The fabric cover influences the air permeability, porosity, wear resistance, projection of threads on the surface of fabric, filtering action, and so on. Knowing the fabric geometry, various problems can be solved, such as designing the fabrics of equal crimps, equal thread inclinations, the minimum area density and the maximum thickness.

It has been assumed that the warp and weft threads have constant diameters. In this condition, for determining the features of the fabric structure it is sufficient to study the arrangement of thread axes in the space of fabric, i.e. not only in the fabric plane but also in the vertical planes. Such parameters of fabric as the maximum possible spacing, yarn crimps, thickness and fabric cover, are dependent on the curvature of axes which can be studied in the sectional views. On the diagram in Fig. 11.1 the plan of the plain weave fabric is shown at (A). Three warp threads 1, 2 and 3 are represented, their axes are indicated by dotted lines O_1, O_2 and O_3, respectively. The axes of three weft threads are marked by y_1, y_2 and y_3. The diameters of warp and weft threads are d_0 and d_y, the spacing of warp threads, i.e. the distance between the axes of adjacent threads is, S_0 and that of weft threads S_y. Studying the plan of fabric at (A), it is possible to calculate the proportion of fabric area covered by the projections of parallel circular threads of one of the systems, i.e. thread <u>relative cover</u>, $e = d/S$, and the maximum density of the threads, which is obtained by setting the threads so closely that they touch one another while crossing the central plane of the fabric. But the configuration of the threads cannot be seen at this view, and therefore, the crimps and the fabric <u>thickness</u> cannot be calculated from this view of the fabric.

相对覆盖度

织物厚度

The sectional diagram is shown at (B) in Fig. 11.1, where the section is cut through the warp at the weft thread 1. The sections of the warp threads are circles with diameters of d_0, and their axes are marked by O_1, O_2 and O_3. The <u>central plane</u> of the fabric is indicated by the horizontal dotted line. In this plain weave fabric, the axis of threads have <u>wave-like shape</u>. The axis of the weft thread 1 at (B) is shown by the wavy dotted line. The axis crosses the central plane of the fabric at A and A_1, and cuts the vertical line, passing through the warp axis O_1, at E. The vertical line crosses the central plane at B and passes through the point M, where the weft thread 1 touches the warp thread 1. The angle of inclination of the weft axis to the central plane of the fabric at A is t_y. The axis wave can be characterized by the height or amplitude, h_y, the length, S_0 and the angle of inclination to the central plane, t_y.

中心平面

屈曲状

In the diagram (C) the section is cut through the weft at the warp thread 2. The axes of the weft threads are placed at the points y_1, y_2 and y_3. The axis of warp thread 2 is shown by the wavy dotted line, which crosses the central plane with the angle t_0. The height of the axis wave, or the maximum displacement of the axis normal to the plane of the fabric, is marked h_0. Comparing the shape of this warp axis with the shape of axis of the weft thread at (B) in the figure we can see the difference in heights of the waves, i.e. h_y is greater than h_0. This indicates the difference in the warp and weft crimps. The weft crimp, c_y, is greater than the warp crimp, c_0.

Studying the sectional views at (B) and (C), it is possible to estimate the maximum theoretical density of threads. The density of warp threads is determined by the distance between the axis of the adjacent threads of O_1 and O_2 at (B). The weft thread 1 with the diameter of d_y passes between them, thus these warp threads are separated by a distance equal to the diameter d_y. The minimum value of O_1 and O_2 is $d_0 + d_y$.

In this case the maximum theoretical density of warp threads

$$P_{0\max} = 1, \ S_{0\min} = 1/\sqrt{(d_0 + d_y)^2 - h_0^2},$$

where h_0 is the vertical displacement of axis of warp threads.

几何结构　　The geometrical model of the plain weave fabric is given in Fig. 11.1 at (D), where the warp and weft threads are represented as circular cylinders of diameters d_0 and d_y respectively. The cross-section cut through the weft thread is shown. The axis of the weft thread is indicated by the wavy dotted line with the length of l_y, which crosses the central horizontal plane of the fabric at an angle t at the point A. The axis contains the straight part and two arcs of the circle of diameter $D = d_0 + d_y$. Two warp adjacent threads are shown, whose axes are marked O_1 and O_2. The angle of inclination of the line $O_1 O_2$ to the central horizontal plane of the fabric is u. The spacing of the warp threads is $O_1 P = S_0$.

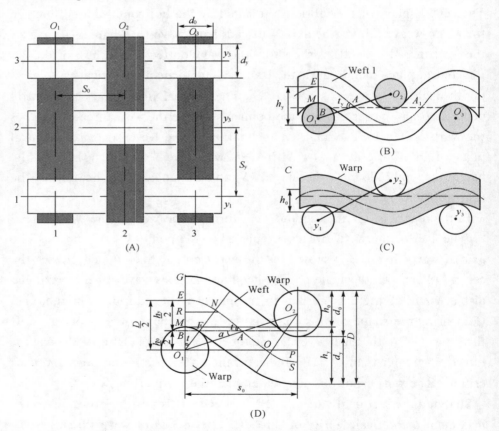

平面和剖面图

Fig. 11.1　Geometrical model of plain weave fabric (plan and sections)

纵向位移
受关注的
屈曲程度

　　An important parameter of the geometrical model of the fabric is a vertical displacement of the parts of the thread. The maximum vertical displacement is considered for the warp as well as for the weft threads. Due to the interlacing of warp and weft threads, the mutual bending takes place. The threads in the fabric have a wave-like shape. The value of bending is characterized by the shape of the axes of the threads when they are in the fabric. The vertical displacement of the weft thread in Fig. 11.1 is marked as h_y, and it can also be called as the height of the weft wave or amplitude. The maximum vertical displacement of the warp thread in this fabric is h_0. There is a certain relation between h_y and h_0. The warp displacement, h_0, decreases with an increase of the weft displacement, h_y, and vice versa. The sum of the warp and weft displacement is constant for the given fabric and equals the sum of thread diameters.

纬屈曲波高

屈曲波高之和等于经纬纱直径之和

$$h_0 + h_y = d_0 + d_y = D \quad \text{or} \quad h_0 = D - h_y.$$

As can be seen in Fig. 11.1, a mutual position of the warp and weft threads in the fabric can be characterized by the value of the phase of fabric construction, which is calculated as a ratio of the warp vertical displacement and the sum of the yarn diameters:

织物结构相

$$F = h_0/D \quad \text{or} \quad F = 1 - h_y/D.$$

The value of phase varies from 0 to 1. A variety of different phases can be studied within this range. To simplify the calculations, it was suggested by Professor N. G. Novikov to consider only nine mutual positions of threads in the square set fabric(Fig. 11.2). The total vertical displacement of the warp thread is divided into 8 parts in this case, because half of yarn radius is assumed as a unit. In the first phase the warp threads are straight in the fabric, therefore, the vertical displacement of the warp is zero, $h_0 = 0$, and the value of phase is $F = h_0/D = 0$. The vertical displacement of the weft threads in this fabric is or the a maximum possible,

平衡结构

第一结构相

纵向位移

$$h_{y(max)} = D - h_0 = D.$$

By using the second formula we have

$$F = 1 - h_y/D = 1 - D/D = 0.$$

In the fabric of another construction of the last phase the weft threads are straight and the bending of the warp relative to the weft is maximum. In this case the weft vertical displacement equals zero, $h_y = 0$, and hence, the warp displacement is D, $h_0 = D - h_y = D$.

第九结构相

Using either the first or the second formula, we have

$$F = h_0/D = D/D = 1 \quad \text{or} \quad F = 1 - h_y/D = 1 - 0/D = 1.$$

The value of phase equals 1.

A variety of constructions can be considered between these two. In one of the constructions the warp vertical displacement equals that of weft, i.e.

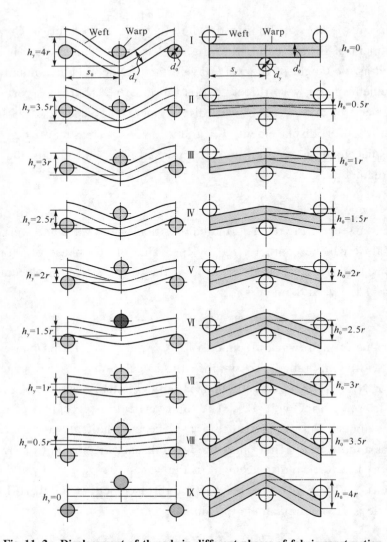

Fig. 11.2 Displacement of threads in different phases of fabric construction

$h_0 = h_y$ and in this case $F = 0.5$.

In the fabric with constant yarn densities and setts, the following parameters of fabric depend on the value of phase:

— the warp and weft <u>crimps</u>, C_0 and C_y;

— the distance between the axes of adjacent warp and weft threads, K_0 and K_y;

— <u>the maximum densities of warp and weft threads</u>, $P_{0(max)}$ and $P_{y(max)}$;

— the warp and weft <u>relative covers</u>, e_0 and e_y;

— the <u>angle of inclination</u> of warp and weft threads to the central horizontal plane of the fabric, t_0 and t_y;

— the angle of inclination of the line connected with the axes of warp and weft threads, to the central horizontal plane of the fabric, u_0 and u_y;

—the thickness of the fabric;
—the characteristics of the fabric surface.

The wave-like shape of threads in the fabric can be estimated by <u>crimp</u>. 织缩
The crimp of the weft thread shown in Fig. 11.2 is calculated as

$$C_y = \frac{l_y - S_0}{S_0}.$$

This and any other relations remain equally valid on consistent interchange of the suffix 0 and y. Doing this in the above formula, we obtain the formula for calculating the crimp of warp thread

$$C_0 = \frac{l_0 - S_Y}{S_Y}.$$

It is clear that the crimp depends on the length of the thread axis. Then, the length of axis of the weft thread depends on its vertical displacement or the height of weft wave. Sometimes in finding the weft crimp, it is advisable to determine the ratio l_y/S_0. The crimp can be calculated by subtracting 1 from the result obtained.

An important parameter of geometrical model of the fabric is the distance between the axes of adjacent threads in the plane of the thread of the other system. In Fig. 11.1 (D) the axes of the warp threads are marked as O_1 and O_2, and the distance between them $O_1O_2 = k$. The distance O_1O_2, where $O_1P = S_0$ is the spacing of the warp, and $PO_2 = h_0$ is the vertical displacement of the warp thread axis. So, we have

$$k_0 = \sqrt{S_0^2 + h_0^2}.$$

This distance cannot be less than the sum of the diameters, $d_0 + d_y$. Thus, $\underline{K_{0(min)}} = d_0 + d_y$, and in this case the space between the adjacent warp threads is equal to the diameter of the weft thread, and the threads touch each other. This condition determines <u>the maximum possible number of warp threads</u> per cm, 最大经纱密度

$$P_{0(max)} = 1/S_{0(min)}.$$

But,
$$S_{0(min)} = \sqrt{D^2 - h_0^2}.$$

Thus,
$$P_{0(max)} = 1/\sqrt{D^2 - h_0^2}.$$

Studying the cross-section in Fig. 11.1 (D) the value of the warp relative cover, e_0, can be found as the ratio of the warp diameter, d_0, to the spacing of warp threads, S_0. The maximum possible warp cover for the fabric of given vertical displacement of warp, h_0, can be found as

最大覆盖系数

$$e_{0(max)} = \frac{d_0}{S_{0(min)}} = d_0/\sqrt{(d_0 + d_y)^2 - h_0^2}.$$

The weft thread in Fig. 11.1 (D) crosses the central plane of the fabric at an $\angle NAB = t$, which can be found from the angle NAO_1. The $\angle NAO_1$ equals the sum of angles $(t + u)$, and the sine of this angle is expressed as

$$\sin NAO_1 = NO_1/AO_1.$$

Hence, $\sin(t + u) = \dfrac{D/2}{k_0/2} = D/k_0$.

It is known that

$$\sin(t + u) = \sin t \times \cos u + \cos t \times \sin u.$$

In Fig. 11.1 (D) we can see that

$$\sin u = \sin BAO_1 = \frac{BO_1}{O_1A} = \frac{h_0/2}{k_0/2} = \frac{h_0}{k_0}.$$

And $\cos u = \cos BAO_1 = \dfrac{BA}{O_1A} = \dfrac{S_0/2}{k_0/2} = \dfrac{S_0}{k_0}$.

Thus, $D/k_0 = S_0/k_0 \text{ s} \times \sin t + \dfrac{h_0}{k_0} \cos t$,

or $S_0 \sin t + h_0 \cos t = D$.

But, $\cos t = \sqrt{1 - \sin^2 t}$.

替代

Now, after the substitution we have

$$S_0 \sin t + h_0 \sqrt{1 - \sin^2 t} = D.$$

代数的

This equation can be solved algebraically for $\sin t$ and only one root is chosen

$$\sin t = \frac{Ds_0 - h_0\sqrt{k_0^2 - D^2}}{k_0^2}.$$

Taking into account that $S_0^2 = k_0^2 - h_0^2$ we have

$$\sin t = \frac{D\sqrt{k_0^2 - h_0^2} - h_0\sqrt{k_0^2 - D^2}}{k_0^2}.$$

Now, the angle t can be found

$$t = \arcsin\left(\frac{D}{k_0^2}\sqrt{k_0^2 - h_0^2} - \frac{h_0}{k_0^2}\sqrt{k^2 - D^2}\right).$$

when the weft thread is straight, $h_y = 0$, and, hence, $h_0 = D - h_y = D$,

$$t = \arcsin\left(\frac{D}{k_0^2}\sqrt{k_0^2 - D^2} - \frac{D}{k_0^2}\sqrt{k_0^2 - D^2}\right) = 0.$$

In this case, the angle of inclination of the weft thread to the central horizontal plane of the fabric equals zero.

Considering the geometrical model in Fig. 11.1 (D), the length of weft thread axis between vertical planes passing through the axes of adjacent warp threads can be determined. The axis contains the straight part *NQ* and two areas *EN* and *QS* of the circle of diameter *D*. Due to the symmetry, it is sufficient to consider only one are *EN* and the part of axis *NA*.

The total length is calculated as

$$l_y = 2NA + 2\text{arc } EN.$$

The distance *NA* can be found from triangle $O_1 NA$

$$O_1A^2 = O_1N^2 + NA^2,$$

and then $NA = \sqrt{O_1A^2 - O_1N^2}$.

But $O_1A = k_0/2$ and $O_1N = D/2$.

Thus, we have $NA = \frac{1}{2}\sqrt{k_0^2 - D^2}$.

The length of arc *EN* is determined, since the angle EO_1N equals the angle $NAF = t$

$$\text{arc } EN = \frac{D}{2}t.$$

Finally, the length is found

$$l_y = \sqrt{k_0^2 - D^2} + Dt.$$

The angle *t* (in <u>radians</u>) in Fig. 11.1(D) 弧度

$$t = \pi/2 - \angle NO_1A - \angle O_2O_1P.$$

But
$$\cos NO_1A = \frac{D/2}{k_0/2} = \frac{D}{k_0},$$

and
$$\cos NO_1A = \sin(\pi/2 - \angle NO_1A) = \frac{D}{k_0}.$$

Thus,
$$\left(\frac{\pi}{2} - \angle NO_1A\right) = \arcsin D/k_0.$$

In Fig. 11.1 (D) we can see that

$$\sin O_2O_1P = h_0/k_0.$$

Hence, $O_2O_1P = \arcsin h_0/k_0$.

Finally, $t = \arcsin D/k_0 - \arcsin \frac{h_0}{k_0^2}$.

The length l_y can be determined as

$$l_y = \sqrt{k_0^2 - D^2} + D \arcsin D/k_0 - D \arcsin h_0/k_0$$

When the weft thread is straight in the fabric, the vertical displacement of the weft thread equals zero, $h_y = 0$. The vertical displacement of the warp thread can be found as $h_0 = D - h_y = D$.

The angle of inclination of the weft thread is zero, $t = 0$. In this case, the length of weft thread is equal to

$$l_y = \sqrt{S_0^2 + h_0^2 - D^2} + Dt = S_0.$$

Thus, if the weft thread lies straight, the length of it is equal to the distance between the axes of adjacent warp threads.

Analogically, if the warp is straight, the length of its axis between the adjacent weft threads equals the spacing between weft threads, S_Y i.e. $l_0 = S_Y$. This is a minimum value of the length.

The length of the thread of one system reaches its maximum when the other system of threads lies perfectly straight with zero crimp.

The thickness of the fabric can be determined from the geometrical model of fabric. The thickness of fabric equals <u>the distance between the horizontal planes</u>, one of which touches the face threads and the other, the threads of the back side of the fabric. The planes touch either the warp threads when the weft is straight or the weft threads when the warp lies straight in the fabric, in general case, the thickness equals the sum of thread diameter and its vertical displacement, $d + h$ (Fig. 11.2). The greatest of two values, i.e. $d_0 + h_0$ or $d_y + h_y$ determines the thickness of the fabric. The condition of the minimum thickness is $d_0 + h_0 = d_y + h_y$. Taking into account the relation between the vertical displacements of threads of different systems, i.e. $h_0 + h_y = d_0 + d_y$, it can be found that the thickness reaches its minimum when $h_0 = d_y$, and $h_y = d_0 + d_y - h_0 = d_0$. The minimum value of thickness is, therefore, $d_0 + d_y$. (Fig. 11.2).

By increasing the value of h_0 to the maximum, which is $h_{0(max)} = d_0 + d_y$, the first maximum thickness is reached, which is $d_y + 2d_0$. In this case, the warp <u>predominates</u> on the face as well as on the wrong side of the fabric.

The second maximum of thickness takes place when the vertical displacement of weft thread, h_y, is increased to the maximum, i e. $h_{y(max)} = d_0 + d_y$. We have the value of thickness $d_0 + 2d_y$. Now, the weft predominates on both sides of the fabric, and the warp lies straight in the central plane of the fabric between the weft threads.

Now, we can see the dependence between the thickness of the fabric and the vertical displacement of the warp. When the vertical displacement is zero, $h_0 = 0$, the fabric thickness is the maximum and equal to $d_0 + 2d_y$. While increasing the vertical displacement of the warp up to $h_0 = d_y$, the thickness decreases and reaches the minimum value, $d_0 + d_y$. Further increasing the warp displacement gives the increase of the thickness. And when the displacement of warp reaches the maximum, i.e. $h_{0(max)} = d_0 + d_y$, the thickness reaches the second maximum, $d_y + 2d_0$.

The appearance of fabric and the structure of fabric surfaces also depend on the value of the warp vertical displacement. When the displacement is from 0 to d_y, the weft predominates on the surfaces. When the displacement varies from d_y to $d_y + d_0$, the warp predominates on the surfaces of the fabric.

The mass of warp and weft yarn in 1m² metre of the fabric depends on the value of warp vertical displacement. The displacement determines the length of thread and its crimp. For example, when the vertical displacement of the warp thread equals zero, $h_0 = 0$, and, therefore, the warp crimp is zero, $C_0 = 0$, the mass of the warp reaches the minimum value. In this case the weft crimp is maximum, and, therefore, the mass of the weft reaches the maximum value. The masses of the warp and weft can be calculated as

重量

$$G_0(h_0 = 0) = P_0 T_0 10^{-2} \quad \text{and} \quad G_y(h_0 = 0) = P_Y T_y(1 + C_y) 10^{-2},$$

where G_0 and G_y are the masses of the warp and the weft in grams;
   P_0 and P_Y are the numbers of warp and weft threads per 10cm;
   T_0 and T_y are the linear densities of warp and weft yarn, tex;
   C_y is the fractional crimp of weft thread.

When the vertical displacement of warp is maximum, $h_0 = max$, the warp crimp is also maximum and the weft crimp equals zero, $C_y = 0$. The mass of threads in 1m² of fabric is

$$G_{0(h_0 = max)} = P_0 T_0 (1 + C_0) 10^{-2} \text{ and}$$
$$G_{y(h_0 = max)} = P_y T_y 10^{-2},$$

where C_0 is the crimp of warp thread.

Thus, while increasing the vertical displacement of warp h_0, the mass of the warp in one square metre of fabric gradually increases and that of the weft gradually decreases.

Example. Calculate the parameters of plain weave fabric with the maximum theoretical setting and the minimum thickness produced of identical yarns and settings in the warp and weft, i.e. $T_0 = T_y$ and $P_0 = P_y$.

In this fabric we have $d_0 = d_y = d$, and $S_0 = S_y = S$. It is known, that the minimum thickness of fabric equals $d_0 + d_y$, and, therefore, the thickness of this fabric is $2d$.

Now, the thread vertical displacement can be found, taking into account that $h_0 + h_y = d_0 + d_y$. Thus, $h_0 + h_y = 2d$. The conditions of the minimum thickness of a fabric are $h_0 = d_y$ and $h_y = d_0$. And, hence, $h_0 = d$ and $h_y = d$, or $h_0 = h_y = h = d$. The vertical displacement of the warp and weft is $h = d$ (Fig. 11.2)

纱线的密度

The distance between the axes of adjacent threads determines the <u>the setting</u> of threads. The number of threads per unit of length reaches the maximum when the space between the threads decreases up to the diameter of the thread of the other system, and the distance reaches the value of $2d$, hence, when $k = 2d$.

The closest possible thread spacing can be found from the equation

$$h^2 + S^2 = k^2.$$

The thread spacing, therefore,

$$S = \sqrt{k^2 - h^2} = \sqrt{4d^2 - d^2} = d\sqrt{3} = 1.732d.$$

And the maximum number of thread per cm can be calculated

$$P_{max} = 1/S = 1/d\sqrt{3} = 0.578/d.$$

The maximum relative cover of the fabric of this construction is determined by the formula

$$e = d/S = \frac{d}{d\sqrt{3}} = 0.578.$$

The warp cover is equal to the weft cover.

The angle of thread inclination to the central fabric plane in this fabric,

$$\begin{aligned} t &= \arcsin D/k - \arcsin h/k \\ &= \arcsin 2d/2d - \arcsin d/2d \\ &= \pi/2 - \pi/6 \\ &= 60°. \end{aligned}$$

The angle of thread inclination of the line, connecting the axes of adjacent threads, to the central fabric plane can be calculated as

$$\sin u = h_0/k_0 = d/2d = 0.5 \quad \text{and} \quad u = \pi/6 = 30°.$$

The angle between the axis of the thread of one system and the line connecting the axes of the threads of the other system in the fabric with theoretically closest setting is equal to the sum of angles t and u,

$$t + u = \pi/3 + \pi/6 = \pi/2 \text{ radians or } 90°.$$

The crimp, for example, of weft thread can be calculated by the formula

$$= \frac{l_y - S_0}{S_0} = \frac{l_y}{S_0} - 1.$$

Here l_y is the length of the weft thread axis which can be found by the formula

$$l_y = \sqrt{k_0^2 - D^2} + Dt,$$

here $t = \pi/3$, $k_0 = 2d$ and $D = 2d$.

Hence, $l_y = \sqrt{(2d)^2 - (2d)^2} + 2d \cdot \dfrac{\pi}{3} = \dfrac{2\pi d}{3}$.

The crimp will have the value

$$C = \frac{2\pi d}{3} / d\sqrt{3} - 1 = 0.21 \text{ or } 21\%.$$

11.2 Fabric cover and cover factor

紧度和紧度系数

One of the main characteristics of fabric is the density of yarns or yarn spacing. But in some cases, in filter fabrics, for example, this characteristic is not sufficient, because the space between the adjacent threads also depends on the yarn thickness. Due to this, the yarn diameter should be taken into consideration. It is known that the fabrics with the same density of threads may have different spaces between the threads because of the difference in diameters. On the contrary, the fabric with different densities of threads may have the same space between the threads when the smaller density is combined with a greater diameter. Therefore, the relative closeness of threads depends on the density of threads and their diameters.

The projected view of the fabric is shown in Fig. 11.3, where the warp spacing is S_0, the weft spacing S_y, the diameter of warp thread d_0 and that of weft d_y. The fractional cover, e, is defined as the fraction of the fabric area covered by the threads, i.e.

投影图

紧度

$$e = \frac{d}{S}.$$

It is common to calculate warp cover and weft cover separately,

经向紧度/纬向紧度

$$e_0 = \frac{d_0}{S_0} \quad \text{and} \quad e_y = \frac{d_y}{S_y}.$$

The cover reaches the maximum value when the threads cover the whole fabric area, i.e. $d = S$, therefore $e = 1$. It gives the scale from 0 to 1.

The warp spacing S_0 gives P_0 threads per unit length

$$P_0 = \frac{1}{S_0},$$

and the number of weft threads per unit length is determined as

$$P_y = \frac{1}{S_y}.$$

The warp and weft covers are

$$e_0 = P_0 d_0 \text{ and } e_y = P_y d_y.$$

The formula can be written as $e = \dfrac{P}{1/d}$, where P is the actual density of threads, $1/d$ is the maximum theoretical density of threads, P_{max} when there is no space between adjacent threads.

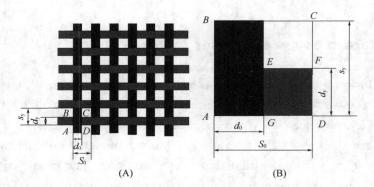

Fig. 11.3 Unit cell of fabric

Now $e = P/P_{max}$, that is, the cover can be expressed as the ratio of the actual density to <u>the maximum theoretical density</u>.

If $P = P_{max}$, the fractional cover equals 1, that means that the threads cover the whole fabric area. In practice e can be greater than one due to the compression of threads.

The cover can be calculated in percentage.

$$E = \frac{d}{S} \times 100.$$

The percentage cover, E, has a scale from 0 to 100 percent.

For determining the fabric cover, one unit cell of the fabric, which is shown at (B) in Fig. 11.3, is studied. The fabric cover is the ratio of fabric area covered by warp and weft threads to the total fabric area. The total fabric area is equal to the area of the rectangle $ABCD$, i.e. $S_0 S_y$. The area covered by threads is equal to the sum of areas of the rectangles ABHG and FEDG, which are $d_0 S_y$ and $d_y(S_0 - d_0)$.

$$e_f = \frac{ABHG + FEDG}{ABCD} = \frac{d_0 S_y + d_y(S_0 - d_0)}{S_0 S_y} = \frac{d_0 S_y + d_y S_0 - d_0 d_y}{S_0 S_y} = \frac{d_0}{S_0} + \frac{d_y}{S_y} - \frac{d_0}{S_0}\frac{d_y}{S_y}.$$

Introducing e_0 and e_y we can express the fractional fabric cover as

$$e_f = e_0 + e_y - e_0 e_y.$$

The fabric cover, e_f, is seldom used due to the following disadvantages: when either e_0 or e_y becomes 1, the fabric cover is 1, in spite of different values of another fraction cover.

Using either the fractional cover or the percentage cover makes it necessary to calculate the yarn diameter. In practice, we usually deal with yarn count or linear density. That is why, it is advisable to introduce these terms and use them in calculations.

In the old system of units the yarn diameter in inches is calculated as

$$d_{(in)} = \frac{1}{28\sqrt{N_e}} \text{ (only for cotton)},$$

where N_e is the cotton count.

Substituting for d in the formula of fractional cover gives

$$e = d/S = pd = \frac{P}{28\sqrt{N_e}}.$$

The ratio of threads per inch to the square root of the cotton count is called the cover factor, K_c,

紧度系数

$$K_c = P/\sqrt{N_e}.$$

There is a certain relationship between e and K,

$$e = K_c/28.$$

Thus, the direct proportional dependence takes place. And when $e_{max} = 1$, $K_{cmax} = 28$. In other words, when the threads are placed with no space between adjacent threads, the cover factor equals 28, i.e. it reaches the maximum value. The cover factor ranges from 0 to 28 in this system of units. To the maximum value of the cover factor depends on the units of P and N_e. The maximum value of K_c equals the number of threads of the first count, $N_e = 1$, per inch.

In the tex system the maximum value of the cover factor differs from that of English system. The diameter of yarn in the tex system, is determined as

$$d(\text{mm}) = \sqrt{T}/26.6 \text{ (only for cotton yarn)},$$

where T is the yarn linear density in g/km, tex.

Developing the formula of fractional cover, we have

$$e = d/S = Pd/10 = P\sqrt{T}/266,$$

where S is the yarn spacing in mm;

d, the yarn diameter in mm;

P, the density of threads per 10mm.

紧度系数　In the tex system the product of threads per cm and the square root of linear density is called the <u>cover factor</u>,

$$K = P\sqrt{T}.$$

The maximum numerical value of K equals the maximum number of threads of the first tex $T = 1$ per cm, $K_{max} = 266$.

The relationship between e and K is

$$e = K/266 \quad \text{or} \quad K = 266e,$$

if the density of yarns in the fabric is 0.91g/cm^3.

Thus, the maximum possible cover factor is $K_{max} = 266$. The cover factor ranges from 0 to 266, when $K = 133$, then $e = 133/266 = 0.5$ and $K = 66.5/266 = 0.25$ or $E = 25''$. In practice K can exceed the theoretical maximum value K, i.e. 266.

For cotton yarns of low twist, when the yarn density can be assumed 0.8 g/cm^3, the relationship between e and K is

$$e = K/250.$$

In this case the cover factor of the fabric ranges from 0 to 250, and with $K = 125$ the percentage cover is 50.

If it is necessary to convert the old cover factor totex system, the conversion factor

$$C = 266/28 = 9.53.$$

And the value in thetex system equals the product of value in the old system and number 9.53.

Example. A cotton fabric of plain weave has the following characteristics: warp 25 tex, 28ends/cm; weft 15 tex, 30picks/cm density of yarn 0.91g/cm^3.

Calculate the warp and weft fractional covers, fabric cover, warp cover factor and weft cover factor.

Warp cover $e_0 = \dfrac{P_0\sqrt{T_0}}{266} = \dfrac{28\sqrt{25}}{266} = 0.526$.

Weft cover $e_y = \dfrac{P_y\sqrt{T_y}}{266} = \dfrac{30\sqrt{15}}{266} = 0.437$.

Fabric corer $e_f = e_0 + e_y - e_0 e_y = 0.526 + 0.437 - 0.526 \times 0.437 = 0.733$.

Warp cover factor $K_0 = P_0\sqrt{T_0} = 28/\sqrt{25} = 140$.

Weft cover factor $K_y = P_y\sqrt{T_y} = 30/\sqrt{15} = 116$.

  It should be remembered that there is a distinction between "cover factor" 区别
and "cover". The former is a conventional measure of the closeness of setting of the threads running in one direction. The later signifies the actual efficiency of the yarns in closing up the cloth. The cover of a cloth may be judged by the appearance of the cloth when held up against the light, and it depends not only on the number of threads per cm and their linear density but also on their regularity, hairiness, fiber composition, twist, and the cloth finishing processes. Any irregularity in construction, as for example in the uniformity of the spacing of the threads, tends to reduce the lever of cover. "cover factor" is calculated from only two of these quantities and, therefore, can't provide a complete indication of "cover".

  Cover factor is, however, useful in making comparisons. Between cloths, particularly those from similar yarns, not necessarily of the same linear density, but in the same weave. With experience it enable a prejudgment to be made, from a knowledge of cloth particulars, of aspects of cloth quality such as handle and air porosity, and it helps the manufacturer to judge the ease or difficulty of weaving. The common weaves, such as plain, twill, and matt, cannot have a high cover factor in both warp and weft directions simultaneously. This is a fundamental aspect of the geometry of these weaves. It is possible to attain higher cover factors in cloths with few intersections, such as satin and twill weaves than in plain weave.

  The study of the geometry of woven cloth is largely mathematical and is built on basic assumptions relating to the shape of the cross-section of the yarn as it lies in the cloth, the profile of the crimp of the yarn, the compressibility of the yarn, and its retention of an imposed "set". Various assumptions may be made, and yield somewhat different results. The geometrical relationships of the characteristics of cloths studied in this way are helpful in understanding of cloth properties and behavior but have not reached a satisfactory degree of precision.

HOMEWORKS

A cotton fabric of plain weave has the following characteristics: warp 15 tex, 50ends/cm; weft 25 tex, 25picks/cm, density of yarn 0.91g/cm^3. Calculate the warp and weft fractional covers, fabric cover, warp cover factor and weft cover factor.

Chapter Twelve
Target Design and Materials Selection
产品定位设计与原料选择

Design work plays a vital role in a company or a factory. A product should be good in three aspects: functions, aesthetics, and benefits for the company. Design work includes the following. 功能/艺术/效益

12.1 Target design 定位设计

12.1.1 Target design
There are numerous requirements of fabrics. In order to select a proper one, the following question words need to be considered.

Where: where the consumers come from, urban or rural area, hot or cold place? 地方

When: is the fabric used for winter or summer? 季节

Who: who will wear the cloth, man or woman; adults or children, rich or poor? 消费者

What: what does the fabric used for, out-wear fabric or under-wear fabric. Suits or leisure? 种类

Why: why does the people wear that cloth, mainly for functions or aesthetics or to present rich? 消费心理

Target design is very important, and is a hard work. It is need experienced designer to perform it. This work also need the designer to study not only the market, but also the factory themselves and competitors. The right target selected, the half success achieved.

12.1.2 Determine the fabric character and properties 确定产品特性
Different end-use product needs different character of fabric. The explanations are as following.

(1) Apparel fabrics Apparel cloths can be divided into out-wear, under-wear, working cloths. They are required health, safety, adequate durability. Of course, they should be aesthetic which plays an aimportant part nowadays. 服装

• Out-wear fabrics 外衣织物
In durability: need adequate strength including tensile, tearing and burst- 拉伸/撕裂

顶破
折皱恢复性
色牢度/比色

ing strength, resistance to abrasion.

In aesthetic acceptability: need adequate <u>crease recovery</u>, easy care properties, lustre, appearance retention including <u>colour fastness</u>, and <u>colour matching</u>.

In comfort: need permeability to air and moisture vapour, stiffness and smoothness.

重量轻

Out-wear for summer, should be <u>light in weight</u>, thin. The materials should be cotton, line, wool, silk and man-made fibres. For winter, the fabric should be thick, <u>dimension stability</u>, the polyester plays an important part.

尺寸稳定

• Under-wear fabrics

Because the fabric touches the body directly, the soft, comfort are very important. In this field, the cotton plays a major role.

• Working cloths

These fabrics need good strength. Some fabrics need special finishing such as: <u>flame resistance</u>, <u>waterproofness</u>, resistance to acid, alkalis and some industrial solvents.

阻燃/防水

家用装饰织物/窗帘
悬垂性

（2）<u>Upholstery fabrics</u>

• <u>Curtains</u>

Curtains need good <u>drape ability</u>, also good aesthetic properties, better colour fastness to light. Man-made fiber plays an important part in this field.

• Covers of furniture

雅致

These fabrics need good abrasion resistance. The colour should <u>delicate</u>. Various thickness and structures are used for this field. Polyester, woolen fabrics play major roles in this field.

Table 12.1 Shows the requirements of various kinds of fabrics.

各种织物服用要求

Table 12.1　Requirements of varies kinds of fabrics

Requirements		out-wear	onder-wear	Working cloths
Aesthetic	Drape ability	☆	☆	△
	Colour	☆	△	△
	Luster	☆	△	△
Handle	Stiffness	☆	☆	☆
	Crease recovery	☆	△	△
	Stretch and recovery	☆	×	☆
	Wrinkle	☆	△	△
Hygienic condition	Permeability to air and moisture vapour	☆	☆	☆
	Thermal insulation	△	☆	△
	Water absorbability	△	☆	☆
	Static	×	×	×
	Weight	☆	☆	☆
<u>Antibial property</u>	Anti-mildew property	☆	△	△
	Moth repellency	☆	×	×

美观性

手感

卫生要求

抗生物性

(Table 12.1 continued)

Requirements		out-wear	under-wear	Working cloths	
Chemical and physical	Resistance of heat	△	×	☆	抗物理化学性
	Resistance of light	☆	☆	☆	
	Chemical	△	×	☆	
Care ability	Washing	☆	☆	☆	保养性
	Crease stability	☆	△	△	
Mechanical properties	Tensile strength	☆	△	☆	机械性
	Tearing strength	☆	△	☆	
	Resistance of abrasion	☆	☆	☆	

Note: ☆—necessary
△—ordinary
×—unnecessary

12.2 Material selection

原料选择

The fabric properties are mainly depended on the fibers. There are numerous fibers to be selected. To be a fabric designer, you must be very familiar with every fibers. As the new man-made fibers have been invented, various kinds of new product are developed. Blended fabrics take the advantages and overcome the disadvantages of the materials. For example, viscose, in wet condition, its strength is very low, but mixed with other fiber which has a good wet strength, the fabric strength have been improved substantially, and it also keeps the viscose's good properties, such as soft handle, water absorptivity, comfortable. Wool blended with acrylic, not only keeps warmth, but also improves the stability. Cotton mixed with polyester, is widely used for shirt fabrics, which has good crease resistance and good moisture absorbability. Some fiber's wear behaviours are listed below at Table 12.2.

吸湿性

Table 12.2 Properties of varies kinds of fiber

各种纤维性能

Properties	Cotton	Polyester	Acrylic	Nylon	Acetate	Viscose	Wool	
Elastic	△	△	○	△	○	△	☆	弹性
Bulk	△	☆	☆	☆	△	△	☆	膨松性
Crease recovery (humidity 65%)	○	☆	○	△	○	△	☆	
Crease recovery (humidity 90%)	△	☆	△	△	△	×	○	
Pleat stability (humidity 65%)	△	☆	☆	○	△	△	○	
Pleat stability (humidity 95%)	×	☆	☆	×	×	×	×	褶裥保持性
Demensional stability	△	☆	☆	☆	☆	×	○	尺寸稳定性
Tearing strength	△	☆	△	☆	×	×	△	
Abrasion resistance	△	☆	△	☆	×	×	○	
Absorbency	☆	×	△	△	☆	☆	☆	吸湿性

Note: ☆—excellent
○—OK
△—ordinary
×—bad

Chapter Thirteen
Calculation and Selection of Varied Parameters
工艺参数的选择与计算

透气性

The fabric is designed for a particular end use, and should possess certain properties, such as strength, flexibility, porosity, wear resistance, fineness, and many others. These properties are determined to a great extent by the structure of fabric, which is characterized by the fabric setting, the yarn properties, the weave, the fabric cover, the yarn crimp percentage, the mass per unit area of the fabric, and so on.

组织选择

13.1 Selection of weaves

Until now in this book, we have adequate knowleges about the woven structure. Some examples of weave selection are shown below.

床单/衬衣类

(1) Sheet, shirt fabric: we want the fabric have adequate strength, thin, dementional stability, etc. The plain weave is suitable for those requirement. If the strength is not taken into consideration, the comfort is necessary, the $\frac{2}{2}$ twill weave, $\frac{2}{2}$ weft rib weave may be selected.

大衣类

(2) Coats cloth: we want the fabric to be thick, worm. Double or backed weave are preferable. If we want the fabric have pile surface, we have to se-

起毛

lect the pile weaves or weaves which have longer weft float for raising.

(3) Curtains: those fabrics need to be soft hand, good drapebility. The satin/sateen weaves are meet those requirement.

(4) Aesthetic pattern cloths: we want the fabrics to have some patterns. The twill weaves derivatives are easy to form different patterns.

纱线线密度的确定

13.2 Yarn linear density

For producing fabrics, thinner or thicker yarns can be used, depending on the fabric end use. In the tex system the thickness of yarn is specified by the linear density. The linear density of yarn in tex is the mass in grams of 1 kilometre of yarn. Linear densities of yarn in the fabric are often given in pairs, warp × weft, $Tt_0 \times Tt_y$, where Tt_0 is the warp linear density, and Tt_y is the weft linear density.

In calculations, the diameter of yarn is often used. In this case, the yarn is considered as a cylinder with a circular cross-section of diameter d. If the length of the cylinder is l, its volume is calculated as $v = \pi d^2 l/4$. To find the mass of yarn, m, the second assumption should be done concerning the density of yarn, q. The density of yarn is determined by the density of the fibre and the percentage of air in the yarn, which depends on a number of factors and is assumed usually as 40%—45%. For cotton yarn of low or medium twist the density is assumed 0.8g/cm³, and for high twist 0.9g/cm³, the mass of yarn of the length l is $m = v \times q$. From the definition of the yarn linear density we have $Tt = m/l$, but $m = vq$ and $l = 4v/\pi d^2$, and hence, the diameter of yarn can be calculated as follows, in mm:

纱的比重

$$d = \sqrt{\frac{4Tt}{\pi q 1000}}.$$

The factor of 1000 is introduced because tex is defined in terms of g/km.

When the yarn twist is low, $q = 0.8$g/cm³, and the diameter in yarn is calculated by the formula

$$d = 0.04\sqrt{Tt}.$$

where the coefficient 0.04 is equal to the diameter of yarn of this particular type, when $q = 0.8$g/cm³, and the linear density, $Tt = 1$ tex.

The air content in yarn can be expressed as a fraction of total volume of yarn, a. The amount of air space depends on the yarn twist. In increasing the twist of yarn the amount of air decreases.

It was proposed by Peirce to assume the percentage of air space in cotton to be 40, i.e. for cotton yarn $n = 0.4$. When the density of substance of the fibre, f, is given, the yarn density, q, can be calculated, $q = f(1 - a)$. In this case the yarn diameter

$$d = 0.0357\sqrt{Tt/q}.$$

The density of cotton $f = 1.52$g/cm³, and assuming $a = 0.4$, it is possible to calculate the cotton yarn density as $q = 1.52(1 - 0.4) = 0.91$g/cm³, when the yarn twist is high. In this case, the diameter of yarn is calculated by the formula

$$d = 0.0357\sqrt{1.1 \times Tt} = 0.0374\sqrt{Tt} = \sqrt{Tt}/26.6.$$

This formula can be used for yarns with specific volume 1.1cm³/g. The specific volume is reciprocal of yarn density and is determined as $1/q = 1/0.91 = 1.1$cm³/g. When either the fibre density or air content in the yarn is changed, the formula should be refined.

Sometimes, the thickness of yarn is specified as a metric count, Nm, which is the yarn length in metres that weighs 1g. In this case, the formula for calculating the yarn diameter in mm is

$$d = 1.18/\sqrt{N_m}.$$

In the old English system the thickness of yarn is specified in the cotton count, Ne, which is the number of hanks of 840 yards in length that weigh 1 pound. The yarn diameter in inches is calculated, in this case, by the formula

$$d_{(in)} = 1/28\sqrt{N_e}.$$

Taking into account that $N_e = \dfrac{591}{Tt}$ and 1 inch = 25.4mm, we have

$$d_{(mm)} = \dfrac{25.4\sqrt{Tt}}{28\sqrt{591}} = \sqrt{Tt}/26.6$$

in theory, and we can calculate the yarn linear density by above formula. The steps are: thickness of fabric $\xrightarrow{\text{fabric geometry}}$ diameter (d) of yarn $\xrightarrow{\text{formula}}$ yarn linear density.

In practice, we select the yarn linear density according to the similar fabric. Then we modify it.

13.3 Fabric setting

The term "sett" is used to indicate the spacing of threads in cloth. Usually, the sett is the number of threads per 1cm, or per 10cm. It is determined for grey as well as for finished cloth. Any cloth is characterized by the number of warp threads (or ends) and the number of weft threads (or picks) per cm. The terms "warp density" and "weft density" are also used. The density can be also expressed by the distance between the axis of the adjacent threads, i. e. thread spacing, S, which is the reciprocal of the sett and can be found by the formula

$$S = \dfrac{100}{P},$$

where P is the number of threads per 100mm, and S is the distance from centre to centre of the threads in mm.

Cloth sett is usually given in pairs, warp×weft, $P_0 \times P_y$, where P_0 is the warp density, and P_y is the weft density.

In the so-called square-sett or balanced fabrics, $P_0 = P_y$.

In designing a new fabric, it is common to calculate the maximal possible density which can be achieved for the given warp and weft yarns.

The theoretical maximum density is obtained when there is no space between the adjacent threads. In reality the density of threads, at least in one of the systems, can be greater than the theoretical maximum due to compression and bending of threads.

In practice, the fabric density are determined like these.

(1) According to the thickness, calculate or select the yarn linear density.

(2) According to the fabric properties required, select the fabric cover factor. (Reference cover factor are given at the appendix tables). Calculate the fabric density according to formula

$$e = P\sqrt{Tt}/266 \quad \text{or} \quad K = P\sqrt{Tt}.$$

For instance:

Design a 3/1 drill: select the linear density 32×32. Referring the hand book or appendix tables select the cover $e_0 = 88$, $e_y = 46$ (for drill the warp cover factor is approximately two of the weft cover factor).

So, we calculate the fabric density.

$$P_0 = \frac{266 \times 88}{\sqrt{Tt}} = \frac{266 \times 88}{\sqrt{32}} = 42.3 \text{ends/cm} = 423 \text{ends/10cm}.$$

$$P_0 = \frac{266 \times 46}{\sqrt{Tt}} = \frac{266 \times 46}{\sqrt{32}} = 22.4 \text{ends/cm} = 224 \text{ends/10cm}.$$

13.4 Crimp of yarn in woven fabric 织物缩率

Fabric is produced by interlacing of warp and weft threads. Interlacing causes the bending of the threads round each other. Due to this, the warp and weft threads have a wavy shape in the fabric. The <u>wavy shape</u> of threads 屈曲结构 can be estimated either by crimp or by take-up.

<u>Crimp, C</u>, is calculated by expressing the difference between the straight- 纱与布之差 ened thread length, L, and sample length, S, as a percentage of 对布长之比 sample length

$$C = \frac{L - S}{S} \times 100\%.$$

Crimp shows the excess of thread length because of curvature of the thread.

<u>Take-up, t</u>, is calculated by expressing the difference ($L - S$) as a per- 纱与布之差 centage of straight or non-woven length of yarn, and it shows the loss in the 对纱长之比 length of the thread in weaving.

$$t = \frac{L - S}{L} \times 100\%.$$

The wavy shape of the thread can also be estimated by the ratio L/S.

Usually the take-up is determined in the preparatory department of a weaving factory for calculating the greater length of yarn necessary for producing a certain length of fabric, i.e. the amount of warp and weft to be ordered for making a fabric of a given length.

Crimp of warp and weft threads should be measured in woven fabric. Crimp of yarn in a particular fabric depends on the sett and the yarn linear density. The ratio of warp and weft crimps is important, because of its great influence on the fabric properties, such as strength, elongation, and on the ratio of a certain property in warp and weft direction. To get the same strength, for example, in warp and weft directions in the fabric of square sett, the crimps should be balanced by controlling the tension of warp on the loom. Sometimes, it is required to produce the fabric with different crimps in warp and weft. The increase of crimp in one direction of the fabric reduces it in another direction. Some changes of the crimp ratio are possible not only in weaving, but also in finishing. The width of fabric decreases and the weft crimp increases when the fabric is stretched in the warp direction. The changes are considerable for worsted fabrics due to another cause, i.e. the felting. In some cases the width of worsted fabric can be decreased up to 30%, due to shrinkage in width.

There is a close relation between the ratio of crimps and the thickness of fabric. By changing this ratio, the mutual displacement of the warp and weft threads, normal to the plane of fabric, takes place. Due to this, either warp or weft floats can be produced on both sides of the plain weave fabric.

There are lot of factors which can effect the crimp or take-up, main of these are summarized as following.

(1) Materials: usually, easy strain plastic fiber has a lower crimp, flexible fiber has a higher crimp.

(2) Woven structure: the more interlace, the higher crimp.

(3) Linear density: when warp and weft have different count. The thicker yarn has a lower crimp. When warp and weft are the same count, they have a higher crimp.

(4) Fabric density: there is a close relationship between fabric density and crimp. In a certain extent, if the structure and warp density fixed, increase the weft density, the warp crimp will increase.

(5) The parameters of the weaving: the warp tension on the loom, the height of back rest, the time of the shedding, etc. all affect the crimps.

So many factors are needed to be considered. It is a very complicated issue. Practically, we determine the crimp according the hand book which gives the crimps or take up of the similar products. See Table 13.1.

手册

Table 13.1 Reference of the take-up of some cotton fabrics

部分棉织物缩率参考

Name	Take-up/%	
	Warp	Weft
Coase plain	7.0—12.5	5.5—8.0
Plain	5.0—8.6	Around 7
Fine plain	3.5—13.0	5—7
Single poplin	7.5—16.5	1.5—4.0
Semi-single poplin	10.5—16.0	1—4
Plied poplin	10—12	Around 2
Single twill	3.5—10	4.5—7.5
Semi-single twill	7—12	Around 2.5
Single serge	5—6	6—7
Single gabardine	Around 10	1.5—3.5
Plied gabardine	Around 10	Around 5
Singe drill	8—11	Around 4
Plied drill	8.5—14.0	Around 2
Corduroy	4—8	6—7
Crepe	6.5	5.5

粗平布
平布
细平布
纱府绸
半线府绸
线府绸
纱斜纹
半线斜纹
纱吡叽
纱华达呢
线华达呢
纱卡其
线卡其
灯芯绒
绉布

13.5 Twist direction and the twill prominence

捻向与斜纹的明显性

The prominence of the twill line in the fabric is dependent on the direction of the twist in the warp and weft yarns, in relation to the direction of the twill. If the twist direction in both yarns is opposite to that of the twill then the twill line will be prominent, but if they are in the same direction, the twill line will be indistinct. If only one of the yarn is twisted in the opposite direction to that of the twill, only the twill line created by that yarn will stand out.

明显
模糊

A $\frac{2}{2}\nearrow$ twill is taken for example.

The effect of using all of the twist combinations for this weave is given below. For a Z twill fabric.

S twist warp × Z twist weft = prominent twill
S twist warp × S twist weft = warp twill prominent
Z twist warp × Z twist weft = weft twill prominent

$$\text{Z twist warp} \times \text{S twist weft} = \text{twill indistinct}$$

This phenomenon is based on a "reflect belt theory". The theory is that the direction of the light reflecting belt of fibers is opposite with yarn twist direction. See Fig. 13.1. At (A) in Fig. 13.1, the reflect areas are linked smoothly, and give us a prominent twill line. At (B) in Fig. 13.1, the twill line is relatively indistinct.

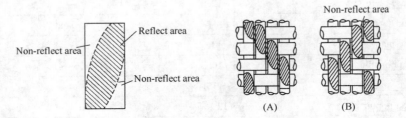

Fig. 13.1　Twist/twill interaction

13.6　Selvedges

A proper selvedges not only give the fabric a good appearance, but also provide a good condition for finishing. The width of selvedges are approximately 0.5—1cm. The weaves of selvedge we often used are explained as following.

13.6.1　Coarse sheeting fabrics
Weft ribs are preferable for the selvedges of these fabrics. In order to avoid the breaking, the selvedge ends are often doubled to draw in.

13.6.2　poplin
The ground weave is used as selvedges weave. Sometimes, we reduce the denting to avoid the tendency of over-density of the selvedges.

13.6.3　$\frac{2}{1}$ twill
The ground weave is suitable for the selvedges.

13.6.4　$\frac{2}{2}$ twill for garbardine, drill
$\frac{2}{2}$ basket weave is widely used for these kinds of fabric. $\frac{2}{2}$ twill in opposite twill line to the ground twill is also used for these fabrics.

13.6.5 $\frac{3}{1}$ drill

$\frac{2}{2}$ basket weave is preferable as selvedge weave.

13.6.6 Satin/sateen

We should pay attention about the warp tension which should be approximately same. That means the interlace in each repeat should be near, i.e. $\frac{5}{2}$ sateen weave with $\frac{2}{2}$ basket.

Chapter Fourteen
Example of Fabric Design
织物设计举例

14.1 Grey fabric design and calculation

坯布设计和计算

These fabrics include: plain, poplin, twill, serge, gabercord, drill, satin-drill, sateendrill, dimity, cotton flannel, etc.

The steps of design are as following.

14.1.1 Width design

幅宽设计

There are international and national standard based on the maximum usage. Sometimes, the width is decided according to customers requirement.

Here, it lists some standard width often used in China for cotton.

Middle-width(cm): 81.5, 86.5, 89, 91.5, 94, 96.5, 98, 99, 101.5, 104, 106.5, 122.

Wide-width(cm): 127, 132, 137, 142, 150, 162.5, 167.5.

14.1.2 Piece length design

匹长设计

Usually, one piece is 30m, and 3—4 pieces are linked together. One point, these are two lengths, <u>nominal length</u> (design length), and <u>standard length</u> (sale length).

设计长/标准长

Because during the process, the fibers are strained the fabrics are longer. After some time, the fabric will recover to approximately 99% of its original length. So, the design length should be longer than sale length.

14.1.3 Structure

Different fabric needs different structure, we have to learn it previously.

14.1.4 Linear density and fabric density (see chapter 13.2).

14.1.5 Crimps (see chapter 13.4).

14.1.6 Total ends calculation

总经根数

Total ends = warp density × width

$$+ \text{ends of selvedges} \left(1 - \frac{\text{ends/dent of fabric body}}{\text{ends/dent of selvedge}}\right).$$

The ends should take the whole number (no fraction).

14.1.7 Dent design   穿筘设计

$$\text{Dent number} = \frac{\text{fabric warp density} \times (1 - \text{weft take-up})}{\text{ends per dent}}$$

Metric dent number = 1.968 English dent number

14.1.8 Width of reed   筘幅计算

$$\text{Width of reed}(\text{cm}) = \frac{\text{total ends}}{\text{ends/dent}} + (5\text{—}10)\text{cm}$$

14.1.9 Calculation of yarn mass per unit area   单位面积重量计算

The mass of per unit area, which is termed "areal density", is an important characteristic of the fabric. The mass in the International system of units can be expressed in grams per square metre. To calculate the areal density of fabric, the following factors should be given: the warp and weft linear densities, setts, and crimps. The areal density is calculated by finding the sum of the mass of warp per square metre and that of weft per square metre. The mass threads can be found per square metre. The mass of threads can be found taking into consideration the crimp or the take-up. When the crimp percentage, C_0, is used, the formula for calculating the mass of warp in grams per square metre is

$$G_0 = TtP_0(1 + 0.01C_0)10^{-2},$$

here Tt is the warp linear density, tex; P_0 is the warp sett, i.e. the number of warp threads per 10cm, ends/10cm.

The factor 10^{-2} is introduced because tex is defined in terms of g/km, and, therefore, the length in m should be converted into km. And P_0 is multiplied by 10 to convert dm into m. So, the resultant factor is 10^{-2}.

The mass of weft per square metre

$$G_y = TtP_y(1 + 0.01C_y)10^{-2},$$

here Tt is the weft linear density, tex; P_y is the weft sett, i.e. the number of weft threads per 10cm, picks/10cm.

The areal density G in g/m² is

$$G = G_0 + G_y = TtP_0(1 + 0.01C_y)10^{-2} + TtP_y(1 + 0.01C_y)10^{-2}.$$

There is a certain relationship between the areal density and the cover factor. The cover factor in the tex system

$$K = P\sqrt{Tt}.$$

This equation can also be exporessed as

$$P = k/\sqrt{Tt}.$$

Substituting for P_0 and P_y in the formula of areal density gives

$$G = K_0\sqrt{Tt_0}(1 + 0.01C_0)10^{-2} + K_y\sqrt{Tt_y}(1 + 0.01C_y)10^{-2}.$$

For the square sett fabric, when $T_0 = Tt_y = Tt$, $K_0 = K_y = K$, we have

$$G = 2K\sqrt{T}(1 + 0.010C)10^{-2}.$$

Example. A cotton fabric of plain weave is characterized by the following parameters: $T_0 = 25$ tex, $P_0 = 28$ ends/cm, $C_0 = 6\%$, and $T_y = 15$ tex, $P_y' = 30$ picks/cm, $C_y = 8\%$. Calculate the areal density of this fabric in g/m².

The mass of warp per square metre

$$G_0 = Tt_0 P_0(1 + 0.01C_0)10^{-2} = 25 \times 280(1 + 0.06)10^{-2} = 74.2.$$

The mass of weft per square metre

$$G_y = Tt_y P_y(1 + 0.01C_y)10^{-2} = 15 \times 300(1 + 0.08)10^{-2} = 48.6.$$

The areal density

$$G = G_0 + G_y = 74.2 + 48.6 = 122.8.$$

Sometimes, it is convenient to know the mass per unit length and use the take-up percentage for calculation (in China).

Now, the formula is

$$G_0 = \frac{MT_0}{(1 - 0.01t_Y) \times 1\,000},$$

Here: M —total ends in the fabric;
Tt_0 —warp linear density;
t_0 —warp take-up percentage.

And for weft is

$$G_0 = \frac{BP_Y T_y}{(1 - 0.01t_0) \times 1\,000}.$$

Here: B —width of grey fabric in cm;
Tt_y —linear density of the weft;
t_y —weft take-up percentage;
P_y —density of the weft.

The mass of one linear meter of grey fabric found as

$$G = G_0 + G_y.$$

14.1.10 Usages of 100 meters of grey fabric

$$G_{100} = \frac{100G}{1\,000(1 - 0.01W)} = \frac{10G}{100 - W},$$

where: G_{100} —mass in kilo of yarn in one linear meter of grey fabric;

W —the amount of wastes of warp/weft yarn(%).

百米用纱量计算

14.2 Example

A crepe fabric, its main parameters are, 100% cotton, linear density $Tt_0 = Tt_y = 29$tex, setts $P_0 \times P_y = 338.5 \times 251.5$ threads/10cm, piece length $30\text{m} \times 3 = 90\text{m}$, width 86.3cm, weave crepe (see Fig. 14.1). The calculation and design are as following.

(1) The take-up selection

See Table 13.1. $t_0 = 6.5\%$, $t_y = 5.5\%$.

(2) Total ends calculation

3 ends/dent for ground, 4 ends/dent for selvedges, and 32 ends for each selvedge(according to experience).

$$\text{Total ends} = 338.5 \times \frac{86.3}{10} + 64\left(1 - \frac{3}{4}\right) = 2\,937 \text{ ends}.$$

2 938 ends are taken for whole repeat.

(3) Dent design

$$\text{Dent} = \frac{338.5}{3}(1 - 5.5\%) \approx 106^{\#}.$$

(4) Width of reed

$$\text{Width of reed} = \frac{2\,938 \times 10}{106 \times 3} + 8(5 \sim 10) \approx 92.46\text{cm}.$$

(5) Fabric areal density

$$G_0 = \frac{2\,938 \times 29}{(1 - 0.01 \times 6.5)1\,000 \times 0.863} = 105.60\text{g},$$

$$G_y = \frac{251.5 \times 29}{(1 - 0.01 \times 5.5)1\,000} = 76.37\text{g},$$

$$G_0 + G_y = 181.97\text{g},$$

(6) The fabric cover

Warp cover $e_0 = P_0\sqrt{Tt_0}/266 = 338.5 \times \sqrt{29} \div 266 = 67\%$.

Weft cover $e_y = P_y\sqrt{Tt_y}/266 = 251.5 \times \sqrt{29} \div 266 = 50\%$.

(7) Yarn length of piece

$$L_p = \frac{30}{1-6.5\%}(1+0.5\%) = 32.2(m),$$

(0.5% is the difference of sale length and design length).

(8) Yarn usages of 100 meters of grey fabric

$$G_{100} = \frac{100 \times 181.97}{1000(1-0.5\%)} \times 0.863 = 15.8(kg),$$

Here: W is 0.5%.

(9) The weaving plan (See Fig. 14.1)

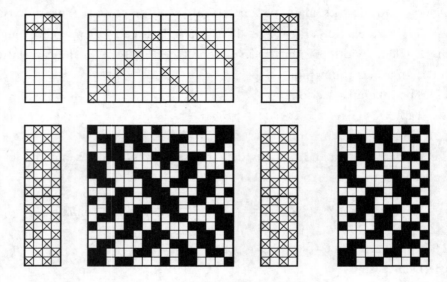

Fig. 14.1 Weaving plan for a crepe

Chapter Fifteen
Glossary of Fabrics
织物生词表

Bedford cord　　Rounded cords run in the warp direction with fine sunken lines between. The cord may be emphasised by wadding ends. The weave on the face of the cord may be plain or twill. Worsted yarns are generally used for suitings and woollen yarns for heavy trousers such as riding breeches, but there is no restriction on the choice of fibres that may be used in this weave. In the lighter weights the construction may be used for dress fabrics. (See Piqué.)　　凸条布

Birdseye　　See colour and weave effects.　　鸟眼组织

Blanket range　　In order to offer customers designs with different colour-ways, a pattern is woven in small sections to form a short length of full-width cloth consisting of variety of designs.　　包袱样

Blazer cloth　　A heavily milled raised woolen cloth, which may be printed with bold stripes. Used for sports coats.　　运动茄克呢

Bouclé　　A woven or knitted fabric with an irregular surface created by the use of fancy yarns having a "bouclé" or "curled" appearance. It may be produced from a wide range of fibres in dress or coating weights.　　珠皮呢

Box cloth　　An all-wool, woolen-spun fabric with a fibrous surface and firm handle. The surface should be completely covered with fibres so that no threads show. It is woven in a variety of weaves, depending on the weight of the finished cloth, and, according to weight, used for such purposes as leggings, coachman cloths, and billiard cloth.　　呢面粗纺呢

Broadcloth　　This term may be used in one of three ways. It may simply refer to (a) suitings which are at least 135cm wide in the finished state (b) light-weight poplin type fabric commonly used as shirting in Canada and the USA, or (c) a heavily milled woollen cloth made in a twill weave from fine merino yarns. The fabric is given a dress-face finish.　　阔幅布、府绸、绒面呢

Brocade　　A fabric ornamented by a pattern produced by Jacquard or dobby weaving, in which warp, weft, or both sets of threads float over the fabric surface to create the required pattern. The basic structure or ground of the cloth is usually a simple weave such as a satin. The woven pattern or figure is often enhanced by the use of continuous filament yarns.　　锦缎

Buckram　　A lightweight woven cloth which is stiffened after piece-dyeing.　　硬衬布

It is usually made from cotton in a coarse plain weave and is used for stiffening.

压辊整理   **Calendered cloth**   This refers to cloth, frequently made from cotton or linen, which has been passed during finishing between pairs of heavy rotating rollers, known as bowls, which may be heated or unheated. Various effects may be produced.

浮雕压花   Embossed   An embossed pattern is obtained by calendering the cloth between a suitably engraved roller and a soft compressible bowl. The embossed effect is durable only if the cloth is made from thermoplastic fibres or, in the case of cotton, has been resin-treated.

定型   Flattened   To flatten and smooth cloth it may be passed between pairs of bowls which apply sufficient pressure to flatten the threads and close up the interstices of the cloth.

压光   Glazed or lustrous appearance   A high lustre may be produced by Schreinering. This is done by impressing very fine lines on to the surface of the cloth from a heated roller engraved with about 150 lines/cm. This finish is not durable unless a thermoplastic material is present. "Everglaze" was an example of such a finish.

波纹   Moiré   A water-mark effect produced by calendering. (See Moiré.)

A lustrous appearance is also effected by friction calendering. Here the cloth passes between a pair of rollers, the upper one having a highly polished, heated surface which rotates at a greater surface speed than the softer bowl beneath. This difference in surface speed develops friction and consequently lustre on the fabric surface.

平布   **Calico**   A generic term for plain cotton cloth heavier than muslins.

细纺   **Cambric**   A fine lightweight, plain-weave cotton or linen cloth which has been fairly closely woven and given a slight stiffening and calendering to produce a smooth surface. Printed, with a crease-resist finish, it is often used for dresses. Very lightweight cambrics (about $65g/m^2$) are used for handkerchiefs.

帆布   **Canvas**   This firm, rather stiff, strong warp-faced cloth is usually made in a closely woven plain or double-end plain weave from cotton, flax, nylon, or polyester. Its weight can be varied over a very wide range according to its intended use.

毯子   **Carpet**   *Velour* or *Velvet* are terms applied to Axminster, Wilton, tufted and bonded carpets which have very smooth, level and velvet-like surfaces. The individual tufts are not visible in the densely packed pile.

Plush or Saxony are cut pile carpets with longer and denser pile than velour. Pile height is up to 15mm. The tufts in the pile appear distinctly separate from each other.

Shag is a tufted or woven long pile carpet with a low tuft density. *Cord* has a short loop pile produced by tufting, weaving, or bonding *Twist pile* denotes a high folding twist in the yarn used in tufting and weaving, giving a slightly mottled look to the surface.

Embossed or sculptured effects are developed by the combination of cut and loop pile.

Casement cloth　　A weft-faced, plain-weave curtain fabric weight about 150g per sq. metre. It may be woven from cotton or manufactured fibres.　　薄窗帘布

Cavalry twill　　A firm warp-faced cloth characterised by steep double-twill lines. The cloth is often made of wool and is produced in a variety of weights to meet the requirements of breeches rainwear, and tailored dresses.　　马裤呢

Chambray　　A plain-weave cotton cloth made with a dyed warp and undyed weft, which gives the cloth a somewhat speckled appearance. Used for dresses.　　青年布

Cheese cloth　　A cheap, soft, plain-cloth of open construction and light in weight. Its principal use is as cheese wrapping.　　包布

Chenille cloth　　A cloth woven with chenille yarn in the weft.　　雪尼尔布

Chiffon　　A sheer, very lightweight plain-weave cloth of open square construction made from hard twisted continuous filament yarns. Silk or nylon are frequently used. The fabric has a very soft drape. It may be piece-dyed or printed.　　绡

Chiné　　A term applied to woven cloth printed with a design having a soft, blurred outline. This is produced by printing the sheet of parallel warp yarns before weaving, with the result that the printed outline does not keep exact register during weaving, and the outline appears blurred in the woven cloth.　　印经平纹布

Ciré　　A term used to describe fabrics with a high mirror-like lustre produced by waxing and polishing the cloth by mechanical means. If a cloth with a satin weave is used, it further enhances the lustre by providing a very smooth surface.　　蜡光织物

Coated fabric　　A knitted, woven or nonwoven fabric on which single or multilayers of a continuous polymeric adherent coating is applied on either one or both faces of the fabric. According to end-use a stiff or flexible coated fabric is produced.　　涂层织物

Colour and weave effects　　Designs in fabric may be produced by the combination of a suitable simple weave and the arrangement of two or more colours in the warp and weft. Fabrics described as Birdseye, dog's tooth, hound's　　配色模纹

tooth, shepherd's check, hairline, step effect, or Glen check pattern are examples of cloths produced in this way.

灯芯绒   **Corduroy**   A cut-weft pile fabric in which the pile forms cords running along the length of the cloth. It is generally made from cotton and the pile may be printed. A velveteen may be cut in such a way as to produce the appearance of corduroy.

芝麻呢   **Covert cloth**   A warp-faced cloth with a fine, steep twill. Its chief characteristic is the mottled or speckled appearance produced by the use of grandrelle or mock grandrelle yarns in the warp only. Worsted yarns are often used, although a quality with a worsted warp and a woollen weft is made. Covert cloth is particularly used for light overcoats.

绉布   **Crépe**   Fabrics in this group are characterised by having a surface which is crinkled or puckered to some degree due to the inclusion of crépe yarn. Such an effect may be produced either in woven or knitted fabrics.

大花型印花装饰布   **Cretonne**   A printed fabric, heavier than chintz, commonly of cotton. It is usually unglazed and likely to carry a floral design. Used for furnishings.

锦缎   **Damask**   A reversible figured fabric woven from one type of warp and one type of weft, based usually on a satin weave. Used for furnishings, and in the bleached state for tablecloths etc.

高级羊毛薄花呢   **Delaine**   A printed lightweight, plain-weave cloth made from wool. It originated from mousseline delaine, which means wool muslin.

劳动布   **Denim**   A warp-faced twill with dyed yarns, generally blue or brown, in the warp and a white weft. It is often made from cotton in a 3/1 twill weave and pre-shrunk during finishing for use in overalls and "denims". Weights range from 200—300g/m^2. Brushed denim and stretch denim are also produced.

"Stonewash" finishing, applied to jeans and denims, involves tumbling garments vigorously in a laundry-type washing machine containing pebbles and possibly bleach. The consequent effect is the now well-recognised non-pristine look, which is often graded by manufacturers to ensure uniformity of effect and quality in the garment.

麻纱   **Dimity**   A fabric, usually of cotton, that is checked or striped by corded effects which are made by weaving two or more threads as one.

小花纹织物   **Dobby**   This is a mechanism applied to the loom that enables weaves and patterns to be produced that cannot be woven on a tappet loom, but are much less elaborate than those obtained by Jacquard weaving. Fabrics so woven are referred to as dobby fabrics.

驼丝锦   **Doeskin cloth**   A fine woollen, warp-face cloth usually of Merino wool, milled, raised, and dress-face finished. It is similar to beaver cloth but lighter and finer.

Dog's tooth check   See colour and weave effects. 犬牙格

Domet   An imitation flannel made mostly from cotton. Both sides of the cloth are raised. It is used as an interlining in tailoring. 绒布

Dotted Swiss   A fine, fairly stiff cotton muslin-type fabric with a clip-spot effect. 点子花薄纱

Dress-face finish   This term applies to woollen cloth covered with a lustrous nap of short fibres. This is largely the result of milling, raising, cropping, and laying the nap, under suitable conditions, smoothly in one direction. The finish is applied to fabrics such as doeskin, beaver, and pilot cloths. 顺毛整理

Drill   A warp-faced twill similar to a denim but usually bleached or piece-dyed. It may be mineral khaki dyed for overalls. "Satin drills" are made in a 5-end satin weave. 卡其布

Duck   A closely woven, strong, plain-weave cloth similar to canvas and used for similar purposes. The term also applies to tropical suitings. 粗布

Duffel   A heavy, low-quality woollen cloth, napped on both faces. Generally made into short "duffel" coats. 起绒粗呢

Dungaree   A strong cotton cloth, similar to denim, made for overalls. A 3/1 or 2/1 twill is used. The cloth may be yarn or piece-dyed. 粗蓝布

Embossed cloth   See calendered cloth. 浮雕压花

Facing silk   A lustrous fabric used for facing lapels in evening suits. Barathea, ottoman, satin, and twill structures are used. Often silk is not used in the cloth in which case it should not be referred to as a "silk" facing. 面子绸料

Faille   A fine, soft fabric, woven from filament yarn, made in a plain weave with weft-way ribs formed by the intersection of a fine close-set warp with a coarser weft. Faille belongs to a group of fabrics having ribs in the weft direction. Examples of this group arranged in ascending order of prominence of the rib are taffeta, faille, poult, and grosgrain. 绨

Felt   The cloth is identified by its densely matted appearance. It may have first been woven before the finish was developed, or it may consist of a mass of animal fibres which have been made to felt or mat together to form a continuous sheet of fabric without the aid of yarns. 毡

Figured fabric   A fabric having a Jacquard or dobby pattern. 花纹织物

Fishnet   An open-work structure, weft-knitted from a combination of floated stitches and plated stitches. Used for run-resistant stockings. 网眼布

Flannel   A plain or twill weave fabric with a soft handle due to being slightly milled and raised. The cloth was originally made entirely from wool but now commonly contains some other fibre also. Both woollen and worsted flannels are produced weighing about $200 g/m^2$. 法兰绒

绒布	**Flannelette**　A cotton imitation of the wool flannel. Softly twisted yarns are used in the weft, and these respond to the action of the raising machine. A nap is produced on both faces of the cloth. Flannelette weighs 180—200g/m² and is simiar to but heavier than winceyette. It may be piece-dyed, printed, or woven from dyed yarns to produce coloured stripes. Flannelette has been used for children's nightwear, but it is legally required to be made flame-retardant. Heavier qualities are used for sheets.
植绒布	**Flock printed cloth**　See printed cloth.
薄软绸	**Foulard**　A lightweight 2/2 twill fabric made from continuous filament flat yarns. It is often printed. It is similar to surah, which was originally made from silk.
起绒粗呢	**Frieze**　A woollen overcoating which has been heavily milled and raised. The nap may be rubbed into small beads or "pills" to produce a "nap frieze".
纬起绒织物	**Fustian**　A general term used to describe a group of fabrics which have a considerably greater number of picks than ends. Beaverteen, corduroy, moleskin cloth, and velveteen are examples.
华达呢	**Gabardine**　A warp-faced cloth firmly woven in 2/1 or 2/2 twill with a greater number of ends than picks. The fairly steep twilllines can be clearly seen, since the yarns used are compact. The fabric is finished to give a clear, clean appearance. Gabardines are commonly made from worsted yarns, all-cotton yarns, unions of wool and cotton, and blends of various fibres. Gabardine to be used for rainwear must be shower-proofed.
纱	**Gauze**　See leno.
乔其纱	**Georgette**　See crépe.
方格色织布	**Gingham**　A firm plain-weave, lightweight cloth of nearly square construction woven with dyed yarns to form a check. Commonly made from cotton; used for dresses, tablecloths, etc.
格纹布	**Glen check or Glen urquhart check**　See colour and weave effects.
坯布	**Grey cloth**　This term is applied to woven and knitted fabrics and is synonymous with "loomstate", which refers to the condition in which the woven cloth leaves the loom.
罗缎	**Grosgrain**　A cloth of about 180g/m² in which the rounded rib runs in the weft direction and is more pronounced than that in a taffeta or poult. The warp consists of closely woven continuous filament yarns. Rather coarse yarn is used in the weft.
仿绸	**Habutai**　A general term applied to silk fabrics that are fine, soft, and have been degummed. Jappe is an exmple of this group.
麻纱	**Haircord**　The cloth bears the name of the modified plain weave from which

it is made. Fine ribs run in the warp direction of the cloth, which may be printed and is usually made from cotton.

Hairline　　See colour and weave effects.　　条纹

Herringbone　　A cloth made from a herringbone weave.　　海力蒙

Hessian　　A plain fabric of approximately square construction woven from one of the bast fibres, usually, jute.　　麻袋布

Holland　　A thin, glazed, medium-weight plain-weave cloth made from cotton or flax which has been beetled or has received a glazed finish. Used for window blind, furniture covering.　　充亚麻窗帘布

Homespun　　Coarse tweeds handwoven from handspun wool yarns in 2/2 twill-weave.　　火姆司本

Honeycomb　　The cloth bears the name of the honeycomb weaves, which are designed to produce a cellular effect. It is made in a range of weights according to its use, e.g, dress fabric or counterpane.　　蜂巢织物

Hopsack　　The cloth of square construction takes its name from the modified plain weave from which it is made. It is synonymous with weaves known as basket or matt.　　板司呢

Hound's tooth check　　See colour and weave effects.　　犬牙格

Jacquard fabric　　A figured fabric with a large scale repeat, e.g. floral brocade which requires a mechanically controlled mechanism, known as a "jacquard", to select and lift the warp whilst weaving.　　提花织物

Jacquard mechanism　　A term used in both weaving and knitting In knitting the mechanism selects the knitting elements in relation to the required pattern.　　提花机构

An electrically controlled jacquard mechanism is now commonly linked directly to a computer which permits increased rates of woven or knitted fabric production and flexibility in design.

Jean　　A warp-faced cotton fabric in 2/1 twill mainly used for overalls.　　三页斜纹布

Lambs' wool　　Wool from the fleeces of lambs, of young sheep up to the stage of weaning, irrespective of breed or type of sheep. The term lambswool is, however, used commercially to indicate a fabric or garment having a soft handle, made totally of virgin wool, a proportion of which is lambs' wool.

Lamé　　Applied to fabrics having flat metallic threads which form either the ground or a decorative figure.　　金银线织物

Lawn　　A lightweight, plain-woven cloth of linen or cotton, of a soft, smooth, and sheer character. Spun yarns made from polyester fibres are also used in this type of fabric.　　上等细布

Leathercloth　　A coated fabric which is embossed to give a leather-like appearance.　　仿皮织物

纱罗	**Leno**	A cellular type of cloth made by crossing warp threads in weaving.
林姆布里克 高级细平布	**Limbric**	A closely woven plain-weave cloth. The softly twisted weft is thicker than the warp yarn, and the picks per centimetre exceed the number of ends. The weft is prominent and has a slight lustre because of its low twist. Used as a dress fabric.
亚麻布	**Linen**	This describes cloths woven from linen yarns spun entirely from flax. Many fabrics with the charactenstic slubby and thready appearance of linen are made from viscose, polyester, and blends. For this reason it is important to state the fibre content of linen-type cloth.
薄纱罗	**Marquisette**	A light-weight leno gauze.
马特拉塞凸纹布	**Matelassé**	A fabric with a quilted appearance produced in weaving. This is achieved by the use of a double or compound structure. It is often used for formal dress wear.
方平	**Matt**	A modified plain weave in which two or more ends and picks weave as one. The fabric may be known as "matt", "hopsack" or "basket weave"。
热熔法非织造布	**Melded fabric**	A fabric made from or containing bi-component fibres. By heating to a suitable temperature one of the component fibres may be softened, producing adhesion.
麦尔登	**Melton**	A heavyweight fabric suitable for overcoats. Lighter weights are used as undercollars in suits. It may be made entirely of wool or with a cotton warp and a woollen weft in a 2/2 twill or other simple weave. It is heavily milled, raised, and cropped.
丝光织物	**Mercerised cloth**	Cotton or linen cloth treated with a cold concentrated solution of caustic soda with or without tension is said to have been mercerised.
美利奴羊毛织物	**Merino wool**	A fine white wool obtained from the merino breed of sheep, or a fabric made from this wool. The term tends to be used rather loosely.
假纱罗织物	**Mock leno**	A cellular type cloth produced from a mock leno weave in which spaces develop between groups of threads.
波纹织物	**Moiré**	A water-mark effect produced on lustrous ribbed or corded cloths by localised flattening of the ribs during finishing. The flattened areas reflect the light differently from the rest of the cloth and consequently produce the distinctive appearance. The finish is generally not resistant to washing. Various styles of moiré can be produced, such as a reversal or mirror-image of the moiré pattern across the cloth and centred on the middle, and a striped moiré produced by suitably engraved rollers.
粗纹棉布	**Moleskin cloth**	This cloth, used for workmen's trousers, is very much like uncut velveteen. It is heavier than a beaverteen with about 140 picks per cm.
麦尔登呢	**Molleton**	A heavy reversible woollen flannel with a nap on both faces. Now

made from other fibres.

Moquette A warp-pile upholstery fabric. The pile may be either cut or uncut, or patterned with cut and uncut pile. 绒头织物

Mousseline (de-soie) A French term describing muslin which was originally made of silk. Now mousseline is applied to similar fabrics made from fibres other than silk. 麦司林真丝薄绸

Mull A very fine, soft, lightweight, plain-weave cloth of fairly open texture and almost square construction. Weight about $50g/m^2$. 麦尔纱

Muslin A general term for soft, fine, plain-weave or simple, leno weave cloths of very lightweight, open, square construction. Muslins for dress wear may be decorated with embroidery, clip-spot, or lappet designs. 平纹细布,麦司林

Nap fabric This refers to woollen overcoating of up to $1,000g/m^2$ which has been rubbed during finishing to produce a beady or pilled surface. 起绒织物

Narrow fabric Sometimes known as smallwares. 窄幅布
 (1) Any textile fabric not exceeding 45cm in width with two selvedges.
 (2) Any trimming.

Needlecord A fine-ribbed corduroy used for dresses. 细条纹光面布/尼农绸

Ninon This may be used to describe a voile fabric made from manufactured fibres. Originally it applied to fine, light open silk cloths with highly twisted yarns woven in groups of two or three in broth warp and weft, and known as double or triple ninon.

Ombré A cloth showing graduated colours or shades produced in weaving, dyeing, or printing. 云纹布

Organdie A lightweight, plain-weave cotton fabric which has been given a durably stiff transparent finish, preferably by treatment with strong sulphuric acid. Used for dresses and stiffening. 蝉翼纱

Organza A sheer, stiff, plain-weave cloth originally made from continuous filament silk in the gum, but now also made from continuous filament manufactured fibres. This fabric may be dyed, printed, or machine embroidered. 透明硬纱

Ottoman A warp-faced cloth with a fairly flat, bold smooth weft-way rib. Originally made from a silk warp and worsted weft. The heavier weight cloths are suitable for tailoring and may have a small fancy design included in the general weft-way rib effects. 粗横棱纹织物

Oxford A good quality shirting fabric made in a plain weave with two ends weaving as one. It is generally made from cotton. Stripes of dyed yarn or decorative weaves are sometimes introduced. 牛津布

Panama A worsted fabric with a clear finish weighing about $200g/m^2$ and used for tropical suiting. The fabric is of a plain weave and square con- 巴拿马薄呢

struction. Fibres other than wool may be used. (N.B. A Panama embroidery canvas of a hopsack weave which is beetled must not be confused with the worsted Panama.)

高级密织薄纱 **Percale** A plain-weave, good quality, closely woven cotton cloth of approximately square construction. It may be glazed in finishing. Used typically for summer dresses and sheets.

匹 **Piece** The unit length of fabric removed from the loom. The standard length depend on the type of fabric, but may be such as best suits the purchaser. Piece lengths are generally between 75 and 100m. According to the state of the cloth expressions such as "loomstate piece", "dyed piece" may be used.

起绒织物 **Pile fabric** A fabric with tufts of fibre or loops of yarn projecting from the surface. The most important classes of pile fabric at present are: (a) woven, in which case it has cut or uncut warp pile, or cut weft pile, (b) knitted, in which case ends of fibres may form the pile or the pile may consist of loops; or (c) tufted, which results in brushed, cut, or uncut pile.

海军呢 **Pilot** A heavily milled and raised woollen cloth generally used for seamen's jackets and overcoats. It is available in a wide range of fibre qualities. A 2/2 twill is often used.

凸纹织物 **Piqué** Originally a woven cloth with rounded cords running in the weft direction, now often made in a lightweight Bedford cord weave with the cord running in the warp direction. Different widths of cords may be produced in the one cloth to create interest, and the fabric may be printed. A piqué effect may be produced by warp or weft knitting.

泡泡纱 **Plissé** See seersucker.

长毛绒织物 **Plush** A cut warp-pile fabric, similar to a velvet but having a longer and less dense pile laid in one direction.

山东府绸 **Pongee** True pongee is woven from wild silk in a plain weave. It is rather lighter in weight and somewhat less irregular in appearance than shantung. It is now also made from some manufactured fibres and cotton. Cotton pongees are mercerised and Schreinered to develop the lustre. Weight about $75g/m^2$.

府绸 **Poplin** A plain weave cloth with fine weft-way ribs. There are twice as many ends as picks. It is frequently made from cotton or blends, preferably with 2-fold yarns. Poplin is available in various weights, making it suitable for shirts or rainwear, depending on the finish applied. Sometimes it is printed of woven with decorative stripes. An Irish poplin used to be made with a silk warp and a worsted weft.

印花织物 **Printed cloth** A cloth on which a coloured pattern has been printed. Printing may be done in the following ways:

<u>Roller</u>   By passing the cloth under an engraved metal roller from which the printing paste is transferred to the cloth to form the pattern. Usually the maximum size of pattern repeat is 46cm. It should be remembered that some printing techniques, not mentioned here, such as resist and discharge printing, involve dyeing the fabric in the piece. 滚筒印花

<u>Transfer</u>   The dye is applied in a suitable medium to paper and can be transferred under heat to the fabric when required. Since all the colours are transferred from the paper simultaneously, registration is easily achieved on knitted fabric. The method also has the advantage that designs can easily be applied to cloth panels. Also known as sublistatic printing. 转移印花

<u>Dye-jet patterning</u>   Thousands of microjets are positioned over the carpet travelling beneath. Each jet is supplied with a constant stream of a single dye which can be activated by computer to inject dye into the carpet to create the required multicoloured design. Pattern change is instant provided the same colours are used. 喷印

<u>Warp printing</u>   Prior to weaving to produce chiné fabrics. 经纱印花

<u>Vigoureux printing</u>   In which bands of colour are printed at intervals across a thick rope of slubbing of fibres prior to spinning. When the slubbing is attenuated during yarn manufacture, a very even blend of dyed and undyed fibre is produced and by this means, if black has been used, a grey yarn results. This is also known as *Melange* printing, and some worsted cloths are known as "melange fabrics". 毛条印花

<u>Flock printing</u>   Involves printing an adhesive in, an appropriate design on to a fabric and then causing finely chopped fibres or flock to fall on the surface. The flock adheres to the printed areas and is removed from the non-printed areas. Metal may also be applied in this way to produce metallic effects. 植绒印花

<u>Burnt out printing</u>   See velvet (brocaded). This technique of distroying only one class of fibre in a fabric made from an appropriate mixture of fibres may be used not only on pile fabrics but on any fabric where a design with a shadow-like effect is requred. 烂花印花

Pure new wool   A description of wool textiles carrying the IWS Wool-mark. A maximum tolerance of 5% is allowed or non-wool fibres used for decorative effects and 0.3% for inadvertent impurities. 纯羊毛

Ratiné   Originally a thick woollen cloth with a curled nap. This term "*ratiné*", the past participle of the French verb *ratiner* (meaning to cover with a curled nap), has also been applied to a cloth, made from a variety of fibres, with a rough surface produced in one of a number of different 结子花呢

ways, either by using a fancy yarn in a cloth to which a special finishing technique, may or may not be applied or by using ordinary yarns in a cloth to which the special finish applied.

丝带 **Ribbon** An attractive woven fabric, characterised in the higher qalities by fine warp yarns and high warp density, and usually of lustrous appearance. Generally for trimming or adornment.

纬缎 **Sateen** A weft-faced cloth made from a sateen weaves, ususally with many more picks than ends. The fabric is often Schreinered to improve the cover and make it lustrous.

经缎 **Satin** A warp-faced cloth made from a satin weave containing many more ends than picks. The fabric is available in various weights and qualities, the heaviest being the Duchesse satin woven on an 8-end repeat. Many satins are made from continuous filament yarns, continuous filament warp and crépe twist weft, cotton, and many other fibres. The smooth, lustrous surface provides a suitable ground for machine embroidery.

萨克森法兰绒 **Saxony** This refers to a superior quality woollen cloth made from fine merino wool (cf Tweed; Cheviot).

稀松窗帘布 **Scrim** A loosely constructed, open light-weight cloth, which may be woven, knitted or otherwise. An important application is in the stablization of nonwoven fabrics.

泡泡纱 **Seersucker** This cloth has interspersed puckered and flat areas of fabric forming striped or check effects. Various methods may be used to produce the effect, which is sometimes known as plissé.

哔叽 **Serge** A piece-dyed 2/2 twill cloth of almost square construction with a clear surface. The twill line runs at a low angle to the weft. It is often made of wool, but other fibres and blends are used.

山东绸 **Shantung** A plain-weave spun silk fabric made from the rather coarse irregular yarn with slubs produced by the wild silkworm (Tussah). Now made from manufactured fibres which imitate the irregularity of Tussah.

雪克斯金细呢 **Sharkskin** May be woven or warp-knitted and in either case is compact and has a firm handle. The cloth often has a dull appearance, which is achieved in the case of manufactured fibres by delustring. Made in dress and suiting weights.

格子花呢 **Shepherd's check** See colour and weave eftects.

闪光呢 **Shot effect** A term applied to such fabrics as "shot taffeta", "shot lining". "shot silk". The effect is produced in fabrics made from lustrous yarns, when the warp and weft yarns are of contrasting colours. The fabric is usually woven in plain or 2/2 twill weave. The colour of the fabric depends on the angle of viewing, and consequently in use, the two colours will appear simultaneously in different areas of the fabric.

Solid worsted　See worsted.　　素色织物

Suede fabric　True suede, made from leather, is produced by abrading the flesh side of the skin to raise a nap and develop a soft, dull effect.　　麂皮织物

Suede-like fabrics are produced in a number of ways:

　Flocking　Electrostatically charged, finely chopped fibres are caused to adhere perpendicularly to an adhesive-coated base fabric.　　静电植绒

　Sueding　One or both sides of a plain or twill-weave fabric is mechanically finished by sueding (abrading) to create a fine short nap. The fabric suffers some mechanical damage in this process. Warp-knit fabrics are sueded on one face only.　　仿麂皮

Taffeta　Taffeta, faille, poult, and grosgrain are all weft-way rib cloths and are listed here in ascending order of prominence of rib. Taffeta is characterised by indistinct weft way ribs, which are the result of using yarns of equal thickness in both warp and weft, and having many more ends than picks. The stiffness of the cloth depends on how closely woven it is, as does the rustling sound it produces when rubbed during wear. Various qualities of taffeta are available, ranging from the lightweight, less stiff fabric used for linings to the closely woven stiff dress taffeta with its tendency to fall into deep folds of a typical character. Wool taffeta is a plain-weave, lightweight fabric produced from worsted yarns.　　塔夫绸

Tapestry　Originally a woollen fabric used in furnishing, particularly wall hangings, having a design in colours produced by inserting relatively short lengths of coloured weft into a uniformly dyed warp, according to the requirements of the design. The term is now used for upholstery fabrics woven on Jacquard looms from coloured yarns. The construction is close and two or more warps and wefts of different materials may be used.　　提花挂毯

Tartan　Originally a woollen cloth of 2/2 twill woven in checks of various colours and worn chiefly by the Scottish Highlandors each clan having its distinct pattern. Other materials and weaves are new used.　　苏格兰格呢

Terry fabric　A looped warp-pile cloth generally made from cotton. It may be colour-woven or printed.

Terry velour　After weaving, the tops of the loops are cut off to produce a soft pile.　　毛巾织物

Tie silk　This is a general term applied to silk fabrics used for neckwear. They are produced in a wide range of designs.　　领带绸

Tricotine　A fine worsted cloth woven in a weave with characteristics of a whipcord.　　巧克丁

Tropical suiting　A fabric such as Panama, weighing about 240g/m^2 used as a　　夏令织物

	lightweight suiting.
簇绒织物	**Tufted fabric**　Produced by passing simple woven fabric through a tufting machine, in which a series of coarse needles with eyes punch continuous lengths of yarn through the cloth to form loops on one face of the cloth. Used in the manufacture of carpets and candlewick.
绢网织物	**Tulle**　A very fine, lightweight net woven from silk yarns in a plain weave. The term is also applied to net with hexagonal mesh made by twisting the threads on a lace machine.
柞蚕丝	**Tussore**　A plain-weave dress weight fabric woven from the coarse silk, known as Tussah, obtained from the wild silkworm. The yarns are generally spun and light-brown or écru in colour.
粗花呢	**Tweed**　Originally a coarse, heavyweight, rough-surfaced wool fabric, for outerwear, made in southern Scotland. The term is now applied to fabrics made in a wide range of weights and qualities from woollen yarns in a variety of weave effects and colour-and-weave effects.
丝绒	**Velour**　A term applied to (1) a heavy pile fabric with the thick pile laid in one direction; or (2) a woven or felt fabric with a raised nap laid in one direction to produce a smooth surface; (3) warp-knit velour produced from long underlaps which are raised and subsequently cropped to produce the cut pile.
经平绒	**Velvet**　A cut warp-pile fabric in which the cut ends of the fibres from the surface. Originally the pile was of silk but now other fibres are utilised.
纬平绒	**Velveteen**　A cut weft-pile fabric in which the cut fibres form the surface of the cloth. It is usually made from cotton and may be dyed or printed.
直贡呢	**Venetian**　An eight-end cotton satin lining generally mercerised and Schreinered. The term Venetian is also applied to an overcoating similar to a Covert but made in a modified satin weave.
骆毛织物	**Vicuna**　A cloth, usually overcoating, made from the fine downy hair of the Peruvian llama.
巴里纱	**Voile**　This fabric is made from hard-spun yarns in a lightweight, open texture. The weave is plain, approximately square. The yarns are cotton, worsted, silk, or manufactured continuous filament.
马裤呢	**Whipcord**　These cloths have prominent steep twill lines formed from the warp threads. There are more ends per centimetre than picks. The cord-like appearance of the twill lines is enhanced by the choice of direction of twist and a clear finish. Whipcord is made in wide range of qualities, usually from cotton and worsted yarns.
粗梳毛纺	**Woollen**　Descriptive of yarns, fabrics, or garments made from yarns spun on the condenser system and containing wool.
精梳毛纺	**Worsted**　Descriptive of yarn in which the fibres are reasonably parallel and

which is spun from combed wool, or of fabric manufactured from such. Additiona terms are used to describe worsted yarns with particular reference to the colour. Sometimes the terms are applied to the fabrics in which the yarns are incorporated.

<u>Solid</u>　All the fibres are of one colour.　　素色

<u>Mixture</u>　Different coloured fibres are blended together.　　混色

<u>Twist</u>　A two-fold yarn consisting rather of single yarns of different solid colours or of different mixture shades.　　双色合股

<u>Marl</u>　A two-fold yarn consisting of identical singles. The single yarn contains two colours obtained by spinning from two different coloured.　　粗纱混色

<u>Melange</u>　Yarn spun from printed slubbing. (See printed cloth.)　　毛条混色

Zephyr　Fine, lightweight cotton fabric used for dresses, shirtings, etc. and ornamented with coloured stripes, checks, and cord.　　细薄织物

Zibeline　A heavily milled and raised woollen coating or costume fabric. The long hairy nap is laid in one direction and pressed flat to give a lustrous satin-like appearance. The inclusion of hairs, such as mohair, enhances the appearance and enables the fabric to retain its appearance rather better in use.　　仿貂皮

Appendix Ⅰ　Fabric Design Table
附录 1　织物设计工艺表

Name		Warp Density(tex)		Width(cm)	
No.		Weft Density(tex)		Mass(m)	
Materials		Linear Density(tex)		Mass/(m^2)	
Spinning					
Warp	Count(tex)		Weft	Count(tex)	
	Twist(t/m)			Twist(T/m)	
	Direction of Twist			Direction of twist	
Weaving and Finishing					
Warp Density (ends/10cm)	Beam		Weft Density (ends/10cm)	Beam	
	Grey			Grey	
	Finished			Finished	
	Beam	Crimp(%)	Gray	Crimp(%)	Finished
Piece Length					
Width					
Weight			Structure		
Warp mass(g/m)					
Weft mass(g/m)					
Mass(g/m)					
Usage(g/100m)					
Note:					

Designer Signature　　　　　　　　**Check Signature**
Manger Signature

Appendix Ⅱ Parameters and Specifications of some Grey Fabrics

附录 2 部分坯织物规格和技术条件

(1) Grey cotton fabric

No. & Name		Width (cm)	Linear density (tex)		Total ends	Fabric density (ends/10cm)		Cover(%)			Dry areal density (g/m²)	Tensile (9.8N/ 5×20cm)		Weave
			warp	weft		warp	weft	warp cover	weft cover	fabric cover		warp	weft	
101	Course plain	91.5	58	58	1668	181	141.5	51.0	39.9	70.6	186.4	67	50	1/1
102		91.5	58	58	1704	185	181	52.1	51.0	76.6	213.1	69	70	1/1
103		96.5	48	48	1758	181	173	46.3	44.2	70.1	169.2	54	53	1/1
104		91.5	48	48	1884	204.5	196.5	52.3	50.3	76.3	192.4	63	63	1/1
105		91.5	48	48	1956	212.5	196.5	54.4	50.3	77.4	199.8	66	63	1/1
106		91.5	42	42	1736	188.5	188.5	45.2	45.2	70.0	156.2	49	51	1/1
107		91.5	42	42	2012	218.5	204.5	52.4	49.0	75.8	177.3	59	57	1/1
108		91.5	42	42	2098	228	204.5	54.7	49.0	76.9	180.4	62	57	1/1
109		91.5	36	36	2098	228	228	50.6	50.6	75.6	162.8	52	55	1/1
110		94	32	32	2230	236	228	49.3	47.6	73.5	147.6	50	51	1/1
111		91.5	32	32	2322	252.5	244	52.7	50.9	76.8	157.2	55	55	1/1
121	plain	122	29	29	2312	188.5	173	37.5	34.4	59.0	103.2	35	31	1/1
122		91.5	29	29	1884	204.5	204.5	40.6	40.6	64.8	115.6	39	40	1/1
123		91.5	29	29	2172	236	220	46.9	43.7	70.2	129.8	46	44	1/1
124		91.5	29	29	2172	236	236	46.9	46.9	71.9	135.0	46	48	1/1
125		91.5	29	29	2336	254	248	50.5	49.3	75.0	144.9	50	52	1/1
126		91.5	29	29	2532	275.5	236	54.8	46.9	76.0	150.9	56	48	1/1
127		91.5	29	29	2532	275.5	267.5	54.8	53.2	78.9	161.4	56	56	1/1
128		91.5	28	28	2172	236	228	46.2	44.6	70.2	128.1	44	44	1/1
129		91.5	28	28	2336	254	248	49.7	48.6	74.2	139.9	48	50	1/1
130		91.5	25	28	2336	254	248	46.9	48.6	72.8	132.0	43	50	1/1
131		96.5	24	24	2290	236	236	42.7	42.7	67.2	111.7	37	39	1/1
151	Fineplain	137	21	19.5	3688	267.5	275.5	45.4	44.9	70.0	105.4	36	38	1/1
152		96.5	19.5	19.5	2598	267.5	236	43.6	38.4	65.3	96.2	34	30	1/1
153		127	19.5	19.5	3422	267.5	267.5	43.6	43.6	68.2	103.3	34	36	1/1
154		98	19.5	19.5	2908	295	295	48	48	73	114.1	38	41	1/1
155		96.5	19.5	19.5	3018	311	318.5	50.6	51.9	76.3	122.3	41	46	1/1
156		96.5	19.5	16	2442	251.5	220	40.9	32.5	60.2	81.1	31	22	1/1
157		96.5	19.5	16	2746	283	271.5	46.1	40.1	67.8	95.8	36	29	1/1
158		122	19.5	14.5	3762	307	307	50	43.2	71.6	102.7	40	29	1/1

(continued)

No. & Name		Width (cm)	Linear density (tex)		Total ends	Fabric density (ends/10cm)		Cover(%)			Dry areal density (g/m²)	Tensile (9.8N/ 5×20cm)		Weave
			warp	weft		warp	weft	warp cover	weft cover	fabric cover		warp	weft	
159	Fineplain	91.5	18	18	2642	287	271.5	45	42.6	68.5	98.1	34	34	1/1
160		91.5	18	18	2880	313	307	49.1	48.1	73.6	110.5	38	40	1/1
161		96.5	18	15	2940	303	251.5	47.5	35.9	66.4	90.5	36	23	1/1
162		98	14.5	14.5	2646	267.5	248	37.7	34.9	59.5	72.5	23	22	1/1
163		96.5	14.5	14.5	3026	311	279.5	43.8	39.4	66	83.1	28	25	1/1
164		96.5	14.5	14.5	3666	377.5	322.5	53.2	45.4	74.5	101.1	35	31	1/1
165		91.5	14	14	2294	248	244	34.2	33.6	56.4	66.1	20	20	1/1
166		99	14	14	3608	362	344	49.9	47.7	73.7	97.5	32	33	1/1
167		96.5	J10	110	3674	377.5	346	44.1	40.4	66.7	70.4	24	23	1/1
168		99	J7.5	J7.5	3928	393.5	362	39.7	36.5	61.8	54.9	18	17	1/1
201	Single poplin	96.5	29	42	3644	377.5	196.5	75.1	47.1	86.9	194.9	79	54	1/1
202		94	29	29	3404	362	196.5	72.0	39.1	83.0	164.1	75	40	1/1
203		96.5	28	28	3152	326.5	188.5	63.9	36.9	77.3	142.5	62	36	1/1
204		96.5	28	28	3416	354	196.5	69.3	38.5	81.2	153.0	71	38	1/1
205		96.5	19.5	19.5	3798	393.5	236	64.1	38.4	77.9	121.3	57	32	1/1
206		96.5	19.5	14.5	3798	393.5	236	64.1	33.2	76.1	110.8	57	22	1/1
207		96.5	16	19.5	4632	480	275.5	71.0	44.9	84.1	131.0	60	40	1/1
208		96.5	16	18	4556	472	275.5	69.8	43.2	82.9	125.2	59	36	1/1
209		98	J16	16	4704	480	275.5	71.0	40.7	82.9	120.2	64	32	1/1
210		96.5	J14.5	J19.5	4936	511.5	263.5	72.1	42.9	84.1	125.7	61	40	1/1
211		96.5	14.5	14.5	4672	484	263.5	68.2	37.1	80.0	106.8	51	25	1/1
212		99	14.5	14.5	5064	511.5	279.5	72.1	39.4	83.1	113.8	56	27	1/1
213		96.5	14.5	14.5	5052	523.5	283	73.8	39.9	84.3	116.4	58	28	1/1
214		96.5	J14.5	J14.5	5280	547	283	77.1	39.9	86.3	120.4	67	30	1/1
231	Semi-single poplin	96.5	14×2	42	3800	393.5	196.5	77.1	47.1	87.9	197.3	85	53	1/1
232		91.5	14×2	21	3166	346	236	67.8	40.1	80.8	147.9	73	33	1/1
251	Plied Poplin	96.5	J10×2	J10×2	4556	472	236	77.8	38.9	86.4	140.7	80	39	1/1
252		91.5	J7.5×2	J7.5×2	4968	543	283	77.6	40.4	86.6	124.1	67	34	1/1
253		91.5	J7.5×2	J7.5×2	5184	566.5	275.5	81.0	39.3	88.4	129.7	69	33	1/1
254		99	J6×2	J6×2	6040	610	299	78.0	38.2	86.4	109.9	60	28	1/1
255		98	J5×2	J5×2	6172	629.5	346	73.6	40.4	84.3	97.0	56	31	1/1

Appendix II Parameters and Specifications of some Grey Fabrics

(continued)

No. & Name		Width (cm)	Linear density (tex)		Total ends	Fabric density (ends/10cm)		Cover(%)			Dry areal density (g/m²)	Tensile (9.8N/ 5×20cm)		Weave
			warp	weft		warp	weft	warp cover	weft cover	fabric cover		warp	weft	
301	Single twill	86.5	32	32	2994	346	236	72.3	49.3	86.0	185.4	81	46	2/1
302		86.5	29	29	2811	325	204.5	64.6	40.6	78.9	149.0	67	36	2/1
303		86.5	28	28	2811	325	220	63.7	43.1	79.4	149.3	64	38	2/1
304		86.5	25	28	3120	360.5	228	66.6	44.6	82.4	150.4	65	40	2/1
305		86.5	24	24	3267	377.5	251.5	68.3	45.5	82.8	151.0	60	38	2/2
401	single serge	86.5	32	32	2684	310	220	64.7	45.9	81.0	164.3	69	42	2/2
402		86.5	29	29	2448	283	248	56.3	49.3	77.9	150.5	53	47	2/2
403		86.5	29	29	2720	314.5	251.5	62.5	50.0	81.3	160.8	63	48	2/2
404		86.5	28	28	2816	325.5	240	63.7	47.0	80.5	153.4	64	43	2/2
405		86.5	28	28	2896	334.5	236	65.5	46.2	81.5	155.3	67	42	2/2
431	Semi-single serge	86.5	14*2	28	2756	318.5	250	62.4	49.0	80.8	155.2	65	44	2/2
501	Single gabar-dine	86.5	32	32	3272	378	236	79.0	49.3	89.3	192.9	91	48	2/2
502		86.5	28	28	3540	409	236	80.1	46.2	89.3	176.2	88	41	2/2
531	Semi-single gabar-dine	86.5	18*2	36	3540	409	204.5	90.7	45.3	95.0	220.0	115	43	2/2
532		81.5	16×2	32	3464	425	236	88.8	49.3	94.4	210.7	105	48	2/2
533		81.5	16×2	32	3548	435	225	90.9	47.0	95.2	209.0	109	45	2/2
534		81.5	16×2	32	3656	448.5	240	93.7	50.1	96.9	219.1	113	49	2/2
535		81.5	14×2	28	3720	456.5	251.5	89.4	49.2	94.7	195.9	103	45	2/2
601	Single drill	98	48	58	3084	314.5	181	80.5	51.0	89.7	252.7	110	65	3/1
602		98	36	48	3700	377.5	188.5	83.8	48.2	91.7	221.8	101	54	3/1
603		86.5	36	36	2952	362	196.5	80.3	43.6	88.9	196.7	94	41	3/1
604		96.5	36	36	3582	370	228	82.1	50.6	91.2	212.2	97	50	3/1
605		86.5	32	32	3540	409	212.5	85.4	44.4	91.9	204.0	103	41	3/1
606		89	32	32	3764	423	224	88.4	46.8	93.9	194.3	109	45	3/1
607		98	29	42	4168	425	236	84.5	56.6	93.3	222.0	98	64	3/1
608		94	29	36	3480	370	236	73.6	52.3	87.5	188.7	79	53	3/1
609		99	29	29	4212	425	228	84.5	45.3	91.6	185.4	98	40	3/1
610		86.5	28	28	3508	405.5	228	79.4	44.6	88.6	168.7	86	39	3/1
611		89	28	28	3784	425	228	83.3	44.6	90.8	178.2	93	39	3/1
612		89	28	28	3784	425	251.5	83.3	49.2	91.6	187.8	93	45	3/1
613		86.5	28	28	3698	427	234	83.6	45.8	91.2	181.9	94	40	2/2

(continued)

No. & Name		Width (cm)	Linear density (tex)		Total ends	Fabric density (ends/10cm)		Cover(%)			Dry areal density (g/m²)	Tensile (9.8N/ 5×20cm)		Weave
			warp	weft		warp	weft	warp cover	weft cover	fabric cover		warp	weft	
631	Semi-single drll	81.5	16×2	32	3804	466.5	214	97.4	44.7	98.6	217.7	119	42	2/2
632		91.5	14×2	28	4456	487	272	95.4	53.3	97.9	213.1	113	50	2/2
633		81.5	14×2	28	4168	511.5	275.5	100.2	53.9	100.1	222.2	121	51	2/2
634		81.5	14×2	28	4428	543	275.5	106.4	53.9	103	231.5	132	51	2/2
651	Piled drill	96.5	19.5×2	19.5×2	4220	437	228	100.9	52.6	100.5	260.5	139	62	3/1
652		98	16×2	24×2	4396	448.5	226	93.7	57.8	97.4	255.6	113	81	3/1
653		81.5	14×2	14×2	3836	470.5	267.5	92.2	52.4	96.1	206.4	107	54	2/2
654		81.5	J14×2	J14×2	4224	518	276	101.5	54.0	100.7	225.5	139	68	2/2
655		96.5	J10×2	J10×2	5928	614	299	101.3	49.3	100.7	182.9	115	46	2/2
656		81.5	J7.5×2	J7.5×2	5528	678	354	96.9	50.6	98.4	157.9	90	40	2/2
701	Single satin drill	98	29	36	4936	503.5	236	100.1	52.3	101.1	225.3	115	47	Satine
702		86.5	29	29	3079	354	240	70.4	47.7	84.8	166.6	74	42	
703		86.5	28	28	3079	354	232	69.3	45.4	83.8	157.5	71	38	
704		86.5	28	28	3238	372.5	267.5	72.9	52.4	87.2	174.6	77	47	
705		89	18	18	4082	456.5	314.5	71.6	49.3	85.7	134.7	61	35	
731	Semi-single satin drill	86.5	14×2	28	3094	354	240	69.3	47.0	83.8	160.8	72	40	
732		86.5	14×2	28	3217	370	267.5	72.5	52.4	87.0	173.4	78	47	
751	sateen	101.5	J14.5	J14.5	3774	370	551	52.1	77.6	89.3	129.2	39	64	Sateen
752		99	J14.5	J14.5	3876	389.5	551	54.9	77.6	89.9	132.3	41	64	
801	dimity	91.5	18	18	2658	289	322.5	45.3	50.6	73	107.1	34	43	1/1
802		99	18	18	2781	279.5	311	43.8	48.8	71.2	101.9	33	41	1/1
901	Plain flanel	106.5	29	58	1688	157	165	31.2	46.5	63.2	137.9	27	60	1/1
902		104	24	44	1732	165	173	29.8	42.9	60.0	114.8	23	47	1/1
903	Serge flanel	96.5	28	36	2376	246	299	48.2	66.3	82.6	174.9	41	65	2/2
904		101.5	24	29	2492	244	259.5	44.1	51.6	73.0	130.6	33	50	2/2

Appendix II Parameters and Specifications of some Grey Fabrics

II (2) Grey (P/C) fabrics

No. & Name		Width (cm)	Linear density (tex)		Total ends	Fabric density (ends/10cm)		Fabric cover (%)			Dry areal density (g/m²)	Tensile (9.8N/5×20cm)		Weave	Blending radio
			warp	weft		warp	weft	warp	weft	fabric cover		weft	warp		
T/C101	Plain	96.5	J21	J21	3014	311	299	52.9	50.8	76.8	131.1	59	59	1/1	65 : 35
T/C102		96.5	J16	J16	3032	312.5	293	46.3	43.4	69.6	100	40	40	1/1	67 : 33
T/C103		98	J16	J16	3524	358	334.5	53	49.5	76.3	110.8	48	47	1/1	65 : 35
T/C104		96.5	J14.5	J14.5	3364	346	354	49.5	50.6	75.1	102.8	41	45	1/1	65 : 35
T/C105		91.5	J14.5	J14.5	3374	366	342.5	51.3	49	75.7	110.1	45	43	1/1	67 : 33
T/C107		96.5	J14.5	J14.5	3822	393.5	342.5	56.3	49	77.7	110.1	49	43	1/1	65 : 35
T/C108		99.0	J14	J14	3802	381.5	299	52.6	41.3	72.2	99.1	45	35	1/1	65 : 35
T/C109	Fine plain	122.0	J13	J13	4246	346	251.5	46	33.4	64	78.7	35	26	1/1	65 : 35
T/C110		99.0	J13	J13	3494	350	283.5	46.6	37.7	66.7	84.6	36	30	1/1	65 : 35
T/C111		96.5	J13	J13	3668	377.5	242.5	50.2	45.6	72.9	96.6	40	38	1/1	65 : 35
T/C112		96.5	J13	J13	3668	377.5	362	50.2	48.1	74.2	96.6	40	40	1/1	65 : 35
T/C113		99	J13	J13	3920	393.5	362	52.3	48.1	75.2	100.6	42	40	1/1	65 : 35
T/C114		120.5	J13	J13	4832	399	331	53.1	44	73.7	97.6	43	36	1/1	65 : 35
T/C115		110.5	J13	J13	5198	433	299	57.6	39.8	74.5	98.6	46	32	1/1	65 : 35
T/C201	Single poplin	96.5	J24	J26	3948	409	240	74	45.4	85.5	169.5	92	60	1/1	65 : 35
T/C202		96.5	J16	J16	4672	484	283	71.6	41.9	83.5	130.4	66	38	1/1	67 : 33
T/C204		95.2	J13	J13	4974	523.5	283	69.6	37.6	81	109.5	57	30	1/1	65 : 35
T/C211	Plied poplin	95.2	J10×2	J10×2	4844	472	275.5	77.9	45.5	88	152	92	58	1/1	65 : 35
T/C212		96.5	J9×2	J9×2	4970	515	256	80.9	40.2	88.6	144.1	121	47	1/1	65 : 35
T/C401	Semi-single gabardine	96.5	J14×2	J28	4484	464.5	236	91	46.3	95.2	205.2	114	50	2/2	65 : 35

(continued)

No. & Name		Width (cm)	Linear density (tex)		Total ends	Fabric density (ends/10cm)		Fabric cover (%)			Dry areal density (g/m²)	Tensile (9.8N/5×20cm)		Weave	Blending radio
			warp	weft		warp	weft	warp	weft	fabric cover		weft	warp		
T/C402	Singledrill	86.5	J13×2	J26	4152	480	244.5	90.7	46.2	95		114	50	2/2	65 : 35
T/C501		96.5	J24	J26	5052	523.5	267.5	94.8	50.6	97.4	201.8	123	58	2/2	65 : 35
T/C502		96.5	J19.5	J19.5	5016	519.5	299	84.7	49.3	92.2	162.9	88	45	2/2	65 : 35
T/C503		96.5	J19.5	J19.5	4932	511	299	83.3	49.3	91.5	170.8	87	45	2/2	65 : 35
T/C511		86.5	J14.5×2	J28	4344	502	275.5	101.9	54	100.9	345	148	65	2/2	65 : 35
T/C513		96.5	J13×2	J28	5135	531.5	275.5	100.5	54	100.2	221.9	137	65	2/2	65 : 35
T/C514	Semi-single drill	96.5	J14×2	J28	5240	543	275.5	106.5	54	102.9	246.2	153	65	2/2	65 : 35
T/C515		95	J13×2	J26	5160	543	280	102.6	52.9	101.2	226.1	140	61	2/2	65 : 35
T/C516		96.5	J11×2	J24	5052	523.5	299	94.8	54.1	97.6	184.6	109	58	2/2	65 : 35
T/C521		95	J10×2	J19.5	5840	614	299	101.3	49.3	100.7		119	47	3/1	65 : 35
T/C531	Plieddrill	95	J10×2	J10×2	5760	606	303	100.0	50	100		118	56	2/2	65 : 35
T/C601		96.5	J9×2	J9×2	6036	625.5	322.5	98.2	50.6	99.1	186.3	107	53	3/1	65 : 35
T/C601	dimity	95.5	J13	J13	3816	398	334.5	52.9	44.5	73.9	95.6	38	34	2/2	65 : 35

Ⅱ(3) Grey blended midfiber fabric

No. & Name		Width (cm)	Linear density (tex)		Total ends	Fabric density (ends/10cm)		Dry areal density (g/m²)	Tensile (9.8N/5×20cm)		Weave	Blending radio	
			warp	weft		warp	weft		weft	warp		T/R	T/A
T/R101	Plain	98.5	21×2	21×2	2180	220	197	182.7	80	78	1/1	65 : 35	
T/R102	Plain	99	18.5×2	18.5×2	2290	230	205	164.5	68	65	1/1	65 : 35	
T/A101	Plain	96.5	18.5×2	18.5×2	2106	217	205	166.7	65	68	1/1		50 : 50
T/A102	Plain	98	16.5×2+33	33		254	220	163.5	61	62	1/1		60 : 40
T/R301	Jacquard fabrics	98	18.5×2	29.5	2786	283	220				Jacquard	65 : 35	

Appendix Ⅱ　Parameters and Specifications of some Grey Fabrics

Ⅱ(4)　References of weaving take-up

Name	Linear density(tex)		Fabric density (ends/10cm)		Take-up(%)	
	warp	weft	warp	weft	warp	weft
Course plain	58	58	181	141.5	11.20	5.44
	48	36	228	232	12.50	6.90
	42	42	188.5	188.5	7.17	8.33
	36	36	228	228	9.70	6.81
	32	32	252.5	244	9.50	6.49
Plain	29	29	188.5	188.5	5.00	8.34
	29	29	236	236	5.00	6.66
	28	28	236	228	7.50	6.66
	28	28	283	259.5	8.60	5.8
	24	24	261.5	237.0	7.60	6.64
Fine plain	19.5	24	283	251.5	8.00	5.55
	19.5	19.5	267.5	236	6.5	5.87
	19.5	19.5	267.5	267.5	7.00	5.88
	19.5	19.5	311	318.5	9.00	6.33
	19.5	16	251.5	220	5.50	5.64
	18	18	244	236	5.4	6.45
	18	18	288.5	314.5	7.00	7.19
	14.5	14.5	354	314.5	7.65	5.00
	14×2	28	297	283	13.0	4.71
	14	14	248	244	3.5	5.09
	J10	J10	283	299	3.8	5.54
	J10	J10	283	283	3.2	5.76
	7	6	420	517.5	5.06	6.25
	7	6	590.5	633.5	9.68	6.66
Single poplin	29	29	326.5	188.5	9.10	3.58
	19.5	14.5	393.5	236	7.5	3.99
	J14.5	J14.5	503.5	220	8.7	2.16
	J14.5	J19.5	511.5	263.5	13.5	1.53
	J14.5	J14.5	523.5	283	11.0	2.21
	J14.5	J14.5	547.0	283	11.3	2.13
	19.5	14.5	393.5	236	8.5	3.99
Plied poplin	14×2	42	393.5	196.5	16.5	1.09
	14×2	29	246	228	14.77	3.41
	14×2	21	246	236	12.0	3.98
Plied poplin	14×2	17	346	259.5	10.35	3.41

(continued)

Name	Linear density(tex)		Fabric density (ends/10cm)		Take-up(%)	
	warp	weft	warp	weft	warp	weft
Plied poplin	10×2	14	543	255.6	16.01	1.45
	J10×2	J10×2	472	236	11	1.66
	J7×2	J7×2	515	295	12.3	2.71
	J7×2	J7×2	566.5	275.5	12.2	2.00
	6×2	6×2	610.0	299	11.34	1.86
	5×2	5×2	643	338.5	9.99	2.01
	4×2	4×2	755	370	11.52	1.96
Single drill (2/1)	32	32	346	236	9.50	4.53
	29	29	325.5	188.5	3.4	4.88
Single drill (3/1)	29	97	310	165	10.30	4.79
	28	28	324.5	212.5	6.5	5.45
	25	28	360.5	228	7.00	5.03
	22	22	396	260	8.0	5.29
	18	18	342.5	421	6.60	7.60
Semi-single drill	18×2	36	287	220	7.00	5.41
	18×2	32	394	252	12.0	5.0
Single serge	29	29	283	251.5	5.51	6.4
	28	28	283	248	5.5	6.95
	28	28	334.5	248	6.0	5.9
	25	28	322.5	251.5	5.85	6.2
	15	14	311	413	4.87	7.01
Semi-single serge	14	28	318.5	250	6.10	5.33
	18	36	354.3	220.4	12.4	3.5
Single gabardine	28	28	484	236	9.73	1.23
	32	32	403	228	11.0	3.7
Semi-single gabardine	18×2	36	416	216.5	11.0	3.0
	16×2	32	435	225	10.0	2.26
	14	29	484	236	10.71	2.86
	14	28	456.5	251.5	9.20	2.17
	14	28	484	286	9.37	2.44
Plied gabardine	16×2	16×2	435	225	10.0	1.40
	14×2	14×2	470.5	267.5	11.0	3.03
Plied drill (3/1)	36	36	362	196.5	8.80	4.34
	36	36	390	216	8.9	2.91
Single drill (3/1)	29	29	425	228	9.20	3.7

Appendix Ⅱ Parameters and Specifications of some Grey Fabrics

(continued)

Name	Linear density(tex)		Fabric density (ends/10cm)		Take-up(%)	
	warp	weft	warp	weft	warp	weft
Single drill	32	32	409	236	11	3.8
Single drill (3/1)	28	28	405	228	8	2.91
	28	28	404.5	252.5	8.64	5.21
Semi-single drill (3/1)	16×2	42	446.5	236	13.17	1.33
Semi-single drill (2/2)	14×2	32	511.3	260.3	12.5	2.53
	14×2	28	481.5	236	8.50	1.95
	14×2	28	543	269	11.5	1.45
	14×2	28	574.6	283	13.5	1.36
	10×2	17	620	297	11.5	2.22
Plied drill (3/1)	16×2	24×2	448.5	226	14.00	1.81
	J14×2	J14×2	481.5	236	8.5	1.95
Plied drill (2/2)	10×2	10×2	598	314.5	12.0	2.70
	10×2	10×2	610	314.5	11	1.93
	J10×2	J10×2	614	299	12	2.57
	J7×2	J7×2	678	354	14.25	2.32
Plied drill (3/1)	J7×2	J7×2	669	338	10.4	2.97
Plied drill (3/2)	10×2	10×2	614	314.5	11.4	2.56
Satin drill	29	36	503.5	236	7	4.68
	28	28	354	232	4	5.41
	14×2	28	354	240	4.80	5.15
	14×2	28	484	236	7.00	2.44
Sateen drill	24	24	251.5	295	2.75	6.24
	14.5	14.5	389.5	551	4.80	5.30
Satinet	18	18	456.5	314.5	7.0	4.30
Dimity	18	18	275.5	314.5	2.00	7.87
Flatwork fabric	29	36	188.5	157	4.38	7.05
Crepe	28	28	337	251.5	6.61	5.48
Double twill drill	14×2	28	560	236	10.1	1.67
Satin drill	14×2	28	580	251.5	6.27	6.35
Serge	14×2	18×2	326.5	307	6.25	7.0
T/C fine plain	15	15	393	342	11.5	5
	13	13	378	342.5	9.0	5.0
T/C poplin	13	13	523.5	283	10.5	2.34
T/C drill	9×2	9×2	625.5	322.5	13.39	1.93
C/V(70/30) plain	29	29	236	220	8.7	6.9
C/C(50/50) fine plain	18	18	313	307	9.0	6.72
Polynosic fine plain	14	14	345	314.5	7.05	4.54

References
参考文献

1. 蔡陛霞. 织物结构与设计. 北京：中国纺织出版社,2003.
2. 姚穆. 纺织材料学. 北京：中国纺织出版社,1997.
3. 沈兰萍. 织物组织与纺织品快速设计. 西安：西北工业大学出版社,2002.
4. 聂建斌. 如何生产适道对路的纺织品. 毛纺织科技,1998.
5. 威灵顿产业用纺织品手册. 北京：中国纺织出版社,2000.
6. Dori's Goerner. Woven Structure and Design. Leeds：British Textile Technology Group, 1989.
7. Blinov, Shibabaw Belay. Design of Woven Fabric. Moscow, 1988.
8. Kathryn L. Hatch. Textiles Science. West Publishing Company, 1993.
9. Amdrea Wynne. Motivate Textiles. London：Macmillian Education Ltd. , 1997.
10. Marjorie A, Taylor. Technology of Textile Properties. . London：Forbes Publication Ltd Ltd. , 1997.
11. BernardP. Corbman Textiles Fiber to Fabric. New York：McGraw-Hill Book Company, 1983.
12. The Textile Institute. Textile Terms and Definition. Manchester, 1988.
13. Jianbin Nie, Shiyan Lu. Fractional Formula Description of Angle-interlock Woven Fabric Construction. Journal of Industrial Textiles. London：SAGA Publications.
14. Jianbin Nie, Shiyan Lu. Design of multi-layer weaves. Proceedin of The 3rd World Conference on 3D Fabrics and Their Applications. Wuhan：World Academic Press.